Proceedings

AF785966

Today, a steadily growing store of information is called for in order to understand the increasingly complex technologies used in modern automobiles. Functions, modes of operation, components and systems are rapidly evolving, while at the same time the latest expertise is disseminated directly from conferences, congresses and symposia to the professional world in ever-faster cycles. This series of proceedings offers rapid access to this information, gathering the specific knowledge needed to keep up with cutting-edge advances in automotive technologies, employing the same systematic approach used at conferences and congresses and presenting it in print (available at Springer.com) and electronic (at SpringerLink and Springer Professional) formats.

The series addresses the needs of automotive engineers, motor design engineers and students looking for the latest expertise in connection with key questions in their field, while professors and instructors working in the areas of automotive and motor design engineering will also find summaries of industry events they weren't able to attend. The proceedings also offer valuable answers to the topical questions that concern assessors, researchers and developmental engineers in the automotive and supplier industry, as well as service providers.

Johannes Liebl
Editor

Grid Integration of Electric Mobility

1st International ATZ Conference 2016

Springer Vieweg

Editor

Dr. Johannes Liebl
Moosburg, Germany

ISSN 2198-7432 ISSN 2198-7440 (electronic)
Proceedings
ISBN 978-3-658-15442-4 ISBN 978-3-658-15443-1 (eBook)
DOI 10.1007/978-3-658-15443-1

Library of Congress Control Number: 2016951243

Springer Vieweg
© Springer FachmedienWiesbaden GmbH 2017
This work is subject to copyright. All rights are reserved by the Publisher, whether the whole or part of the material is concerned, specifically the rights of translation, reprinting, reuse of illustrations, recitation, broadcasting, reproduction on microfilms or in any other physical way, and transmission or information storage and retrieval, electronic adaptation, computer software, or by similar or dissimilar methodology now known or hereafter developed.
The use of general descriptive names, registered names, trademarks, service marks, etc. in this publication does not imply, even in the absence of a specific statement, that such names are exempt from the relevant protective laws and regulations and therefore free for general use.
The publisher, the authors and the editors are safe to assume that the advice and information in this book are believed to be true and accurate at the date of publication. Neither the publisher nor the authors or the editors give a warranty, express or implied, with respect to the material contained herein or for any errors or omissions that may have been made.

Printed on acid-free paper

This Springer Vieweg imprint is published by Springer Nature
The registered company is Springer Fachmedien Wiesbaden GmbH

WELCOME

Electric mobility is one of the most outstanding industrial topics of the present time. In order to give it additional impetus, the charging infrastructure must be further expanded. Furthermore, it is in our interest that the industrial manufacturing of the numerous components of electric mobility takes place in Germany.

Particular focus must be on the development of technologies and concepts for the sensible and user-friendly integration of electric vehicles into intelligent power grids. As the proportion of regenerative energy continues to increase in the future, we will need intelligent solutions to ensure the reliability and stability of the power grid. Achieving this will require a continuous exchange of ideas and opinions between science, industry and politics. This ATZ conference will make a key contribution to this.

I wish you great success for your conference.

Matthias Machnig
State Secretary
Federal Ministry for Economic Affairs and Energy

The UN Climate Change Conference in Paris, with its key topics of global warming and deteriorating air quality, will speed up the advance of electric mobility. CO_2-neutral and zero-emission mobility require electricity to be generated from regenerative sources of energy.

Power generation from wind and solar energy, however, is dependent on the weather and is therefore not stable. The irregularities that occur in nature can result in unacceptable voltage fluctuations in the power grid. For that reason, the availability of highly flexible loads and storage systems is becoming particularly important. Electric vehicles, with their grid-relevant properties as controllable power consumers and electricity storage systems, could help to stabilize future power grids.

A true breakthrough in electric mobility will not occur until the product and the infrastructure are in perfect harmony. In order to optimize the interaction between representatives from the automotive industry, energy providers, data services and politics nationally and internationally, we are offering you this new and interdisciplinary forum in Berlin at the end of May. It will enable you to make direct contact with all interface partners to acquire comprehensive information on current issues and to exchange the very latest knowledge.

We cordially invite you to come to Berlin and we look forward to your active participation.

On behalf of the Program Advisory Board,

Your
Dr. Johannes Liebl
Editor-in-Charge ATZ | MTZ | ATZelektronik

INDEX

KEYNOTE LECTURES

MARKETS

TRAFFIC AND ENERGY SYSTEMS

OVERALL ENERGY SYSTEMS I

KEYNOTE LECTURE

LOCAL ENERGY SYSTEMS

INFORMATION AND COMMUNICATION TECHNOLOGY

SYSTEM BEHAVIOR

OVERALL ENERGY SYSTEMS II

KEYNOTE LECTURE

SPEAKERS

Dr. Gunnar Bärwaldt
Volkswagen AG

Detlef Beister
SMA Solar Technology AG

Dr. Marcus Bollig
BMW Group

Jörn Cohrs
BTC AG

Noel Crisostomo
California Public
Utilities Commission, USA

Qiao Ding
BMW Group, China

Heiko Fastje
EWE NETZ GmbH

Marcus Fendt
The Mobility House GmbH

Dr. Armin Gaul
RWE International SE

Hannes Haupt
Volkswagen AG

Dr. Andreas Heinrich
Daimler AG

Prof. Dr. Willett Kempton
Center for Carbon-free
Power Integration,
University of Delaware, USA

Gero Lücking
LichtBlick SE

Yutaka Ota
Power Systems Design Laboratory,
Tokyo City University, Japan

Xaver Pfab
BMW Group

Dr. Willibald Prestl
BMW Group

Tetyana Raksha
Ludwig-Bölkow-Systemtechnik
GmbH

Florian Samweber
Research Center
for Energy Economics (FfE)

Franziska Schmalfuß
Institute of Psychology, Cognitive
and Engineering Psychology,
TU Chemnitz

Michael Schwaiger
BMW Group

Hans Urban
Schletter GmbH

Prof. Dr. Ulrich Wagner
Institute for Energy Economy
and Application Technology (IfE),
TU Munich

Michael Westerburg
EWE AG

Dr. Holger Wiechmann
EnBW Energie
Baden-Württemberg AG

Ursel Willrett
IAV GmbH

Guiping Zhu
Electrical Engineering Department,
Tsinghua University, China

KEYNOTE LECTURES

Electric mobility – sustainably into the future

Dr. Marcus Bollig, BMW Group

Abstract

Electric mobility has been an integral part of the BMW product strategy since the introduction of the BMW i models i3 and i8 and the expansion of the model range by various plug-in-hybrid vehicles. Customer perspectives are changing worldwide in terms of future mobility and growing importance of sustainability. Therefore, BMW has identified electric mobility as one of the key factors for the future success of the company.

Electrified powertrains provide key characteristics that perfectly fit to BMW:

- Efficiency: Electric motors are the bench mark in terms of efficiency, even having the complete chain consisting of power electronics, the battery and the charging system in mind. Braking energy can be recovered easily. Compared to internal combustion engine powertrains, the amount of energy shrinks to a quarter.
- Dynamics: With the maximum torque at already zero motor speed, electric powertrains are extremely spontaneous in response – pure fun to drive!
- Sustainability: When driving electric, no emissions are exhausted from the vehicle. This is already a great advantage to air quality in today's megacities. And – using renewable power for charging – real total emissions of almost 0 g CO_2/km can be achieved.

There are still challenges to meet on the way to the future mobility: digitalization has arrived in the car, autonomous driving is on its way. Fulfilling the demand for really sustainable mobility remains the most critical though. E-mobility will be the next step, ahead of hydrogen-mobility. Therefore, BMW welcomes efforts to support the rollout of e-mobility with incentives when buying an electric vehicle and doing further investment into the German charging infrastructure. In a third step electric vehicles can bring benefits to the electricity grid by intelligent charging, this can become an additional part in the success of e-mobility. The BMW Group has already conducted a lot of research and development in this field – this will lead to attractive products and services in the future.

The challenge of electric mobility in digitized power grids

Heiko Fastje, EWE NETZ GmbH; I. Kolmsee, EWE AG

This manuscript was not available on completion of this publication. Thank you for your understanding.

MARKETS

Marktintegration von Elektrofahrzeugen – wie SchwarmMobilität® Kaufanreize schafft und die Stromnetze stabilisiert

(The market integration of electric vehicles – how SchwarmMobilität® creates purchasing incentives and stabilizes power grids)

Gero Lücking, LichtBlick SE

Abstract

Es gibt viele Gründe und Belege dafür, dass der Elektromobilität die Zukunft gehören wird. Die Experten der UBS-Bank[1] erwarteten beispielsweise schon im August 2014, dass bis 2020 mit einer Energie-Revolution zu rechnen sei, die auch zu einem Umbruch im Verkehr führen wird. Die Kosten für Lithium-Ionen-Batterien werden sich demnach bis 2020 halbieren. Nach dem Eintritt von Tesla in den Markt stationärer Batterien sind allein im Jahr 2015 die Batteriekosten um 25 Prozent gesunken. Die Reichweite der E-Autos steigt stetig, die Kosten sinken, die Fahrzeuge werden so immer attraktiver und wettbewerbsfähiger. Insbesondere auch in Kombination mit dezentraler Solarenergie, deren Kosten ebenfalls drastisch sinken, werden sie in Zeiten von Diesel-Gate zu einem unverzichtbaren Baustein der weltweiten Energiewende.

LichtBlick entwickelt mit SchwarmEnergie® Lösungen für diese dezentrale und erneuerbare Energiewelt. Produkte und Lösungen im Bereich Elektromobilität sind wesentliche Bausteine dazu.

Im „ersten Schritt" bieten wir Kunden Produkte und Dienstleistungen an, die dazu beitragen, erneuerbare Energien in den Verkehr zu integrieren. Autofahrer sollen in Zukunft immer und überall sauberen LichtBlick-Strom tanken können. Schon heute bieten die regulatorischen Rahmenbedingungen Möglichkeiten, Elektromobilität für die Kunden und Fahrzeugnutzer mit Ökostrom noch preiswerter und damit attraktiver zu machen.

Im „zweiten Schritt" nutzen wir die enorme Batteriekapazität von E-Fahrzeugen als SchwarmSpeicher® für sauberen Strom. Mit der IT-Plattform SchwarmDirigent® kann LichtBlick die Batterien der Elektroautos zu einer virtuellen Großbatterie bündeln und intelligent mit den Energiemärkten vernetzen. So kann bei Stromüberschuss im Netz und negativen Preisen gesteuert geladen werden. Darüber hinaus kann zur Stabilisierung der Stromnetze sogenannte Regelenergie ins Netz ein- und ausgespeist werden. Geschäftsmodelle entstehen, weil mit dieser Marktintegration von Elektrofahrzeugen ein Beitrag zur Versorgungssicherheit geleistet und Geld verdient werden kann.

[1] UBS-Studie: Global Utilities, Autos & Chemicals. Will solar, batteries and electric cars reshape the electricity system? (August 2014)

1 Intelligente Fahrstromprodukte und diskrimierungsfreier Zugang zur öffentlichen Ladeinfrastruktur

Nur Elektroautos, die mit Ökostrom betrieben werden, entlasten das Klima. Die Kosten des Ökostrombezugs können um rund 30 Prozent reduziert werden, wenn eine Regelung im Energiewirtschaftsgesetz (§ 14a EnWG) genutzt wird und Elektroautos ähnlich wie Wärmepumpen und Nachtspeicherheizungen als sogenannte „steuerbare Verbrauchseinrichtungen" behandelt werden. Elektroautos können so an privaten oder gewerblichen Ladesäulen bspw. von abends 21 Uhr bis morgens 6 Uhr vergünstigt Strom beziehen. Für die zeitliche Einschränkung sinken die Kosten für die Netznutzung gegenüber dem üblichen Tarif um etwa 30 Prozent oder um bis zu 200 Euro pro Jahr.

Wird das Betanken am Standort „zu Hause" nicht statisch sondern flexibel gesteuert, kann beim Tanken sogar Geld verdient werden. Am Sonntag, den 8. Mai 2016 wurden beispielsweise acht Stunden lang aufgrund geringer Nachfrage und hoher regenerativer Erzeugung negative Strompreise notiert. Werden die Fahrzeuge in dieser Zeit gesteuert geladen, erhält der Privat- oder Flottenkunde dafür, dass er Überschussstrom dem Netz entnimmt und damit sein Fahrzeug betankt, bares Geld.

Voraussetzung dafür, dass die Kunden flächendeckend von diesen intelligenten Angeboten profitieren können und sie an jeder Ladesäule auch im öffentlichen Raum den Strom des Anbieters ihrer Wahl beziehen können, ist, dass die öffentliche Ladeinfrastruktur Teil der Stromnetze wird und damit allen Stromanbietern und Vertriebsgesellschaften ein diskriminierungsfreier Zugang gewährt wird. Alle anderen Regelungen verzögern den Ausbau der Ladeinfrastruktur und sind europarechtswidrig.

2 SchwarmMobilität® für die Energiewende

Mit der Anschaffung von E-Autos und Solarbatterien bauen die Verbraucher eine gigantische Speicherkapazität auf. Bereits eine Million E-Mobile können mit einer Speicherkapazität von rund 16 Gigawattstunden mehr überschüssigen Wind- und Sonnenstrom aus dem Netz aufnehmen als alle deutschen Pumpspeicher-Kraftwerke. Da ein Auto im Schnitt 23 Stunden am Tag am Standort „zu Hause" steht, kann ein Teil der Batteriekapazität ohne Komfortverlust für den Fahrer dem Stromnetz zur Verfügung gestellt werden. Kurzfristige Schwankungen im Stromnetz können ausgeglichen werden, die Batterien von E-Fahrzeugen können durch das Bereitstellen von sogenannter Regelenergie einen wichtigen Beitrag zur Systemstabilität der Stromnetze leisten. Im Forschungs- und Leuchtturmprojekt der Bundesregierung INEES haben die Projektpartner VW, SMA, das Fraunhofer Institut IWES und LichtBlick gezeigt, dass Elektrofahrzeuge ohne Komforteinbußen für den Nutzer zur Systemstabilisierung der Stromnetze genutzt werden können. Fahrzeugbatterien der zweiten Generation, bidirektionale Schnittstellen, tägliche Ausschreibungen zur Ermittlung der Regelenergiebedarfe, auf vier Stunden verkürzte Zeitscheiben und die Befreiung dieses Systemstroms von Netzentgelten und Umlagen führt zu jährlichen Erlösen von bis zu 1000 Euro pro Fahrzeug. Mit den aktuellen Gesetzgebungsverfahren werden alle genannten regulatorischen Voraussetzungen umgesetzt sein. Und auch die ersten Fahrzeughersteller bieten bereits serienmäßig bidirektionale Schnittstellen an.

Über SchwarmMobilität® etablieren sich so neue Geschäftsmodelle, die die Energiewende befördern und über die Nutzungsdauer der Fahrzeuge eine höhere wirtschaftliche Attraktivität entwickeln als für die Allgemeinheit teure und für die Kunden lediglich einmalig wirkende Kaufprämien.

Case Study – China's regulatory impact on electric mobility development and the effects on power generation and the distribution grid

Guiping Zhu[1], Christoph Tomoki Hein[2], Qiao Ding[2]

1 State key lab of power system, Electrical Engineering Department, Tsinghua University, Beijing, 100084, China
2 Electric and Electronics Engineering Division, BMW China Service Ltd., Beijing, 100027, China

1 Introduction

Since the introduction of "New Energy Vehicles" (NEV) via the Ministry of Industry and Information Technology (MIIT) as a new vehicle type in 2009, several domestic car manufacturers started to offer products in that category (MIIT, 20091). Supported by restriction and subsidy policies, the sales of NEV started to increase rapidly, topping in 200,000 passenger NEV wholesale in 2015 (www.caam.com.cn, 2016). Electric Vehicles (EV) are set as key developing type of NEV in China, thus it is the focus of this paper.

In order to sustain the growth of NEV market in China and to fulfil the massive charging demand of expected 5 million NEV by 2020, more and more charging infrastructure has to be deployed in the cities as well as intercity areas such as high ways (NDRC et al., 2015).

The rapid development of EV has also brought a number of other problems including the impact of EV charging power demand on the grid. At present because of a relatively small number of electric vehicles and relatively huge capacity of power grid, the impact of electric vehicle charging on the power grid is almost negligible, and charging is available for electric vehicles at any time (so called "any-time charging" mode), thereby meeting EV charging demand to the greatest extent. Electric vehicle is a typical power-type load, that is, its charging power is great, however total required electricity is not much (China Automotive Energy Outlook 2012, 2012). Therefore, in case of a sharp increase in the number of electric vehicles and absence of control over their charging time and charging power, that is, any-time charging mode will bring a lot of negative impacts on power grid, such as new load peak and excessive power supply capacity, which is bad for coordinated development of EV industry and power grid. This paper carried out studies amid this context, which used Beijing in China as example to study EV charging impact on power grid with various charging modes.

2 NEV development in China

2.1 Electric mobility in China

Historically speaking, the origin of electric mobility (e-mobility) in China was not only related to passenger vehicles or commercial vehicles, rather to any kind of personal mobility. The ban of motorcycles out of major Chinese cities in the 90s and the lack of alternatives created a new market for personal e-mobility. Electric scooters and delivery three-wheelers were introduced as an alternative, which created a complete new market, which is shown in Figure 2-1.

In concerns of electric passenger and commercial vehicles, the market started to grow with the introduction of restrictions and incentive policies, too.

Delivery type electric three-wheeler

Private electric-scooter (e-scooter)

Figure 2-1 Electric three wheeler and electric scooter in China

Development policies to increase domestic production capabilities of NEV were introduced in 2009 with MIIT announcement (MIIT, 2009[1]). An additional policy stated that all domestic vehicle manufacturers have to certify their companies with a new NEV brand respectively product (MIIT, 2009[2]). The products can apply different technologies, such as plug-in electric hybrid vehicle (PHEV), battery electric vehicle (BEV) or fuel cell electric vehicle (FCEV), to be recognized as a "NEV" in the MIIT NEV catalogue. The first generation of NEV vehicles listed in the MIIT NEV catalogue had not achieved the mentionable sales numbers (www.chinadaily.com.cn, 2014).

2.2 NEV subsidies and restriction policies

The situation continued until governmental and local subsidies for NEV were introduced (MOF et al., 2013) which supports the targets of 5 million NEV in 2020 (SCPRC, 2012). Those targets are part of the ambitious goals of "Made in China 2025" roadmap to strengthen the key industries to become a global leader in innovation and technology. Additionally, restrictions on conventional vehicles push the industry and customer towards NEV offers respectively purchases. Cities like Beijing, Shanghai, Shenzhen restricted the release of number plates for conventional vehicles with a lottery system (BMPG, 2010) with waiting times longer than 3 years or auctions for number plates reaching prices of around 84,500RMB (approx. 11.000EUR) in 2016 for one number plate (SMPG, 2014). On the other hand, in cities like Beijing or Shanghai, number plates for NEV (in Beijing only for BEV) are not restricted and can be acquired within short time free of charge. The maximum possible subsidies and benefits for locally produced NEV and imported BEV/PHEV are respectively shown in Figure 2-2.

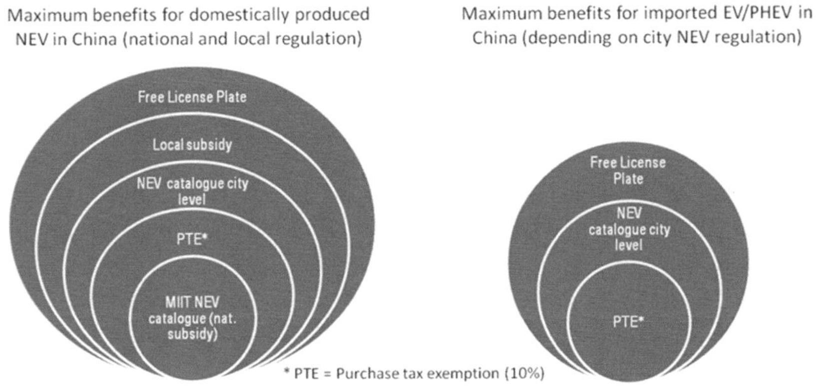

Figure 2-2 Subsidy structure for domestically produced NEV and imported EV/PHEV

The accumulated benefits for a domestically produced BEV and PHEV in Shanghai (MOF et al., 2015; SMPG, 2016) are shown below in Table 2-1 in comparison with a conventional vehicle of the same product class.

Table 2-1: Acquisition cost comparison PHEV, BEV and conventional vehicle in Shanghai 2016 (Exchange rate RMB/EUR = 0.136)

Category	BYD Qin PHEV (NEV >50km range)	BAIC EV200 BEV (NEV >150km range)	Comparable conventional vehicle (BYD F3)
Price before subsidy	209,800RMB (approx. 28,000EUR)	208,900RMB (approx. 28,000EUR)	65,900RMB (approx. 9,000EUR)
Purchase tax	0	0	3,295RMB
National Subsidy	-30,000RMB	-45,000RMB	0
Local Subsidy	-10,000RMB	-30,000RMB	0
Number plate auction	0	0	Approx. 84,500RMB
Total customer costs (after benefits)	169,800RMB (approx. 23,000EUR)	133,900RMB (approx. 18,000EUR)	153,695RMB (approx. 21,000EUR)

2.3 NEV sales increase after 2013

Considering that buyers of conventional vehicles rather need to wait for a long time for a number plate or need to purchase it with high costs, for certain customer classes a NEV is the only solution to acquire personal mobility. A mentionable sales increase can be seen from 2013 to 2016 shown in Figure 2-3. The overall registration of NEV passenger vehicle increased over 300% in 2015 compared to 2014. China is currently the biggest market for BEV and PHEV with around 200,000 wholesale of passenger vehicles in 2015 and will foster a further increase in 2016 (www.caam.com.cn, 2015).

Though governmental subsidies will be successively reduced until 2020, restrictions on conventional vehicles will remain. Those restrictions will still give buyers a certain benefit to buy a NEV comparable with the ban of motorcycles in the 90s.

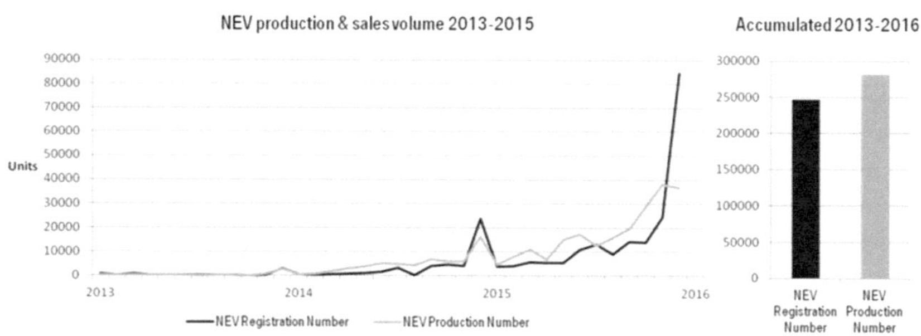

Figure 2-3 NEV production and registration volumes 2013-2015

3 Infrastructure development for NEV market in China

3.1 Overall target of infrastructure development

In order to sustain the growth of NEV market in China and to fulfil the massive charging demand of expected 5 million NEV by 2020, it is essential to enhance charging facility and charging infrastructure development in China. Overall, more than 12,000 centralized charging and swapping stations and more than 4.8 million distributed charging pillars are planned in China by 2020 (NDRC et al., 2015). Thus, the expected ratio of NEV and charging points will approximately be 1:1.

3.1.1 Development plan of centralized charging and swapping stations

Those 12,000 centralized charging and swapping stations are mainly planned in public service areas, in terms of public transportation and e-taxi services, urban sanitation and logistics industries. Meanwhile, urban public charging stations are arranged in societal parking area which could facilitate the occasional charging needs, and intercity fast charging stations are planned for intercity travel needs (NDRC et al., 2015). The expected amount of those centralized charging and swapping stations are listed in Table 3-1.

By the end of 2020, more than 1000 fast charging stations are planned to achieve a fast charging network in major provinces and cities within mainland China (NDRC et al., 2015). The fast charging network focuses to establish intercity travel by using NEV. The intercity fast charging stations as shown in Figure 3-1 are planned along highly utilized highway networks.

Table 3-1 Target amount of centralized charging and swapping stations by 2020

No.	Category of charging and swapping stations	Amount
1	Charging station for public transportation	3,850
2	Charging station for e-taxi services	2,500
3	Charging station for urban sanitation and logistics industries	2,450
4	Urban public charging station	2,400
5	Intercity fast charging station	800
	Overall	12,000

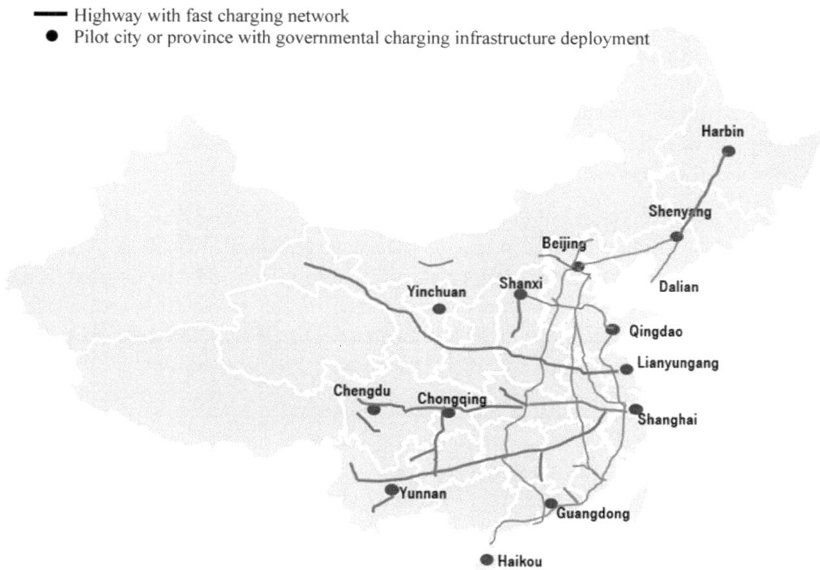

Figure 3-1 Planned intercity fast charging network in China by 2020

3.1.2 Development plan of distributed charging pillars

For passenger NEV usage, more than 2.8 million charging pillars in residential compounds and 1.5 million charging pillars for commercial areas, such as enterprises and institutions, office buildings and industrial parks are planned in China. Those charging pillars will support the essential demands and it is encouraged to share the charging pillars for public use. Besides, more than 0.5 million distributed charging pillars are planned in public parking areas, such as transportation hubs, recreational sports facilities, high rise buildings, large mansions and roadside parking, which aim to meet the occasional charging needs (NDRC et al., 2015).

3.2 Nationwide implementation of charging infrastructure

For the entire country, the Ministry of Housing and Urban-Rural Development of the People's Republic of China (MOHURD) announced that from 2016 on, 100% of parking lots in new-built residential compounds have to install charging devices. For public constructions larger than 20,000m^2 as well as shopping malls, office buildings, etc., at least 10% of the parking lots have to be equipped with charging devices (MOHURD, 2015).

To assure the electric supply to charge NEV in present residential buildings, compounds, etc. with low cost, it is recommended to co-use the power supply of illumination equipment such as street lights. For a short term solution, time-shared charging points are offered to residential compounds with limited parking spaces, meanwhile development of mechanical and multi-storey parking garages with charging facilities are considered as long term goals.

In order to effectively achieve the charging infrastructure development plan by 2020, it is suggested to define provincial targets under consideration of local NEV development levels, consisting of "acceleration", "demonstration" and "promotion" areas (GOSC, 2015). For instance, 12 cites and provinces in the acceleration area, such as Beijing, Shanghai and Guangzhou, have a target of more than 7400 charging stations and 2.5 million charging pillars by 2020 (NDRC et al., 2015).

To keep pace with the targets of the NRDC, Beijing released a set of recommendations and subsidy policies of construction land and investment for deployment of charging stations. It is encouraged to involve public-private partnerships to install charging infrastructure, subsidised with maximum 30% of the total investment (BMCDR, 2015[1]). Concerning the usage, from 2015 on, a charging service fee was introduced in Beijing which is directly related to the daily 92 octane gasoline price (BMCDR, 2015[2]).

The combination of national, provincial and city level infrastructure development targets including public as well as residential charging helps to strengthen the user satisfaction of NEV drivers and further encourages new NEV buyers.

4 Evaluation of EV charging impacts on power grid in China

As electricity demand of BEV is greatest in all kinds of NEV, following research of NEV charging impacts on power grid will only focus on BEV (EV for brief). Massive EV charging will lead to a great charging power demand on power distribution network, as well as impacts on power quality and stability of the grid, therefore EV charging impacts on local power grid should be evaluated by certain method. A model of EV charging power demand was presented firstly in this section. In this model, randomness of charging start and daily mileage of EV was considered, as well as various charging modes, to ensure simulation results close to real use-cases. Using the model, daily charging power curves of a certain number of EV with various charging modes can be obtained. By overlapping those curves with the original power load of the grid, as well as the future EV stock, charging impacts on power grid can be evaluated (Original peak load is defined as the load of the grid without EV).

4.1 EV charging power demand model

4.1.1 Randomness of EV charging

Both time and location of EV charging are random in practice. They vary with driving behaviour, life pattern, and local charging infrastructure distribution and so on. Therefore those aspects cannot be modelled by accurate analytical expressions. A sample survey method was applied in our research to obtain the statistical characteristics of EV charging start time and the corresponding initial State of Charge (SOC). In each city more than 1000 questionnaires were collected mainly focused on following four questions:

(1) What is the distance from your home to working place?
(2) When do you usually go to work from home?
(3) When do you usually arrive at home from your working place?
(4) Do you have a dedicated parking lot? If yes, in your residential area or working place?

According to answers of question (1), daily used electricity of the designated EV can be determined. It is assumed that each EV can be fully charged every day, then the initial SOC SOC_0 of the EV before start of charge can be calculated by following formula:

$$SOC_0 = (SOC_{\text{full}} - \frac{d}{d_{\max}}) \times 100\% \tag{4-1}$$

Where SOC_{full} is set to be 95%, d represents the actual daily mileage of an EV and $d_{\max}$ represents the maximum distance an EV can run when it is fully charged.

According to answers to question (2) and (3), charging start time can be determined for the residential area and working place. Answers to question (4) lead to an approximated proportion allocation of the charging event between residential area and working place. The curves were fitted by applying polynomial expressions to describe the accumulative probability density of daily mileage and arriving time. Figure 4-1 and Figure 4-2 show the real statistical result and the fitted curves of vehicles in Beijing, China. In those two figures, circles represent the real data derived from questionnaires, and the lines are fitting results by polynomials. In order to eliminate the influence of bad data during the investigation and to make the fit function more accurate, the time of home arrivals is limited to after 15:00 (Figure 4-2). The accuracy of results in this way falls within an acceptable range, due to the fact that the original investigated data showed that the vehicles arriving home outside of this time frame were less than 5%.

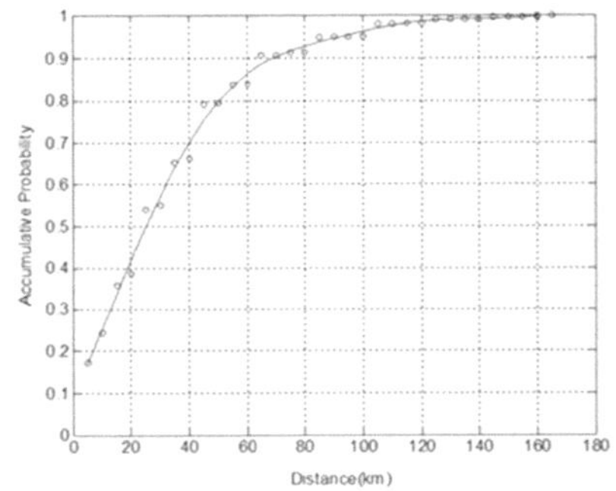

Figure 4-1 Accumulative probability density distribution of average daily mileage of vehicles in Beijing

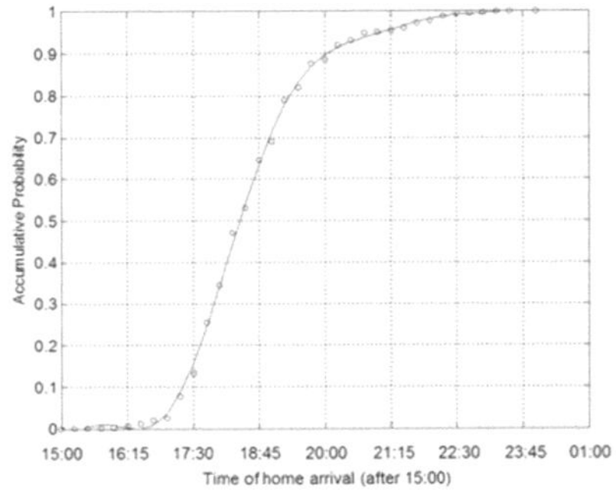

Figure 4-2 Accumulative probability density distribution of home arrival time of vehicles in Beijing

4.1.2 Three charging modes

The EV charging mode has significant influence on real time charging power. Three charging modes were proposed in this research including any-time charging, time-delayed charging and orderly charging.

4.1.2.1 Any-time charging

Any-time charging means that an EV will be charged as soon as it is connected to the power grid at any time, and its charging power is determined by the nominal parameters of the charging point or vehicles' on-board charger or battery. With this charging mode, a good guarantee is provided for the convenient use of EV, however, when the number of EV is large, the impact on the power grid may become serious, usually a new load peak is generated, which will be shown by simulation results in following sections.

4.1.2.2 Time-delayed charging

In order to avoid overlap of EV charging load and original peak load of power grid, time-delayed charging mode is proposed. Since most EVs are not utilized in the evening, they could be delayed for a period of time before starting to charge. From the perspective of the charging duration and the vehicle utilization of the next day, the delay should be appropriate. In this paper, EV which arrive after 5:00pm are generally shifted for 5 hours until charging can start to avoid the original peak load of the power grid.

The biggest advantage of time-delayed charging is that it is not dependent on the two-way communication between power grid and EV. Charging points can realize that function by certain control technologies, and therefore this charging mode can be popularized in current power distribution network. In addition, time-delayed charging can also effectively decrease the peak-valley difference of the grid load, improve the equipment utilization rate of the power grid and delay or even avoid infrastructure upgrades caused by EV charging.

4.1.2.3 Orderly charging

Although it is easy to provide time-delayed charging, it may lead to a very long waiting time which could cause unsatisfactory situations, e.g. emergency use-cases. The ideal situation should be to charge as many EV as possible based on the currently available power grid capacity. This is defined as "orderly charging". This charging mode determines charging start time of EV based on the currently available power capacity, EV arriving sequence and priority of vehicles using by an algorithm based on a two-way communication between power grid and EV. On the one hand this charging mode can guarantee no new higher load peak and on the other hand meet the charging demand of EV by efficiently utilizing the available capacity.

Orderly charging mode is the most ideal charging mode, but the two-way communication between power grid and EV is difficult to be provided in the current power distribution network. With the progressing construction of smart power distribution network and the improvement of relevant EV standards, orderly charging can be expected in the future.

4.1.3 Calculation of EV charging power demand based on Monte Carlo Method

After the accumulative probability density of arriving time and initial SOC are determined, the Monte Carlo Method is applied to generate enough random data of both groups for simulation which follows the derived accumulative probability density (Zhu Guiping, Xing Yuhui, 2013). With these data, charging power demand of a certain number of EV can be calculated, furthermore, its impact on power grid can be evaluated. The schematic of the algorithm is shown as Figure 4-3.

For the sake of quick simulation, the number of random data generated for each group is 10000. When the EV stock exceeds 10000, simulation results are linearly magnified. The error between these linearly magnified results and those simulated by Monte Carlo method with real number is less than 2%, which is acceptable for this study, while simulation time was significantly reduced.

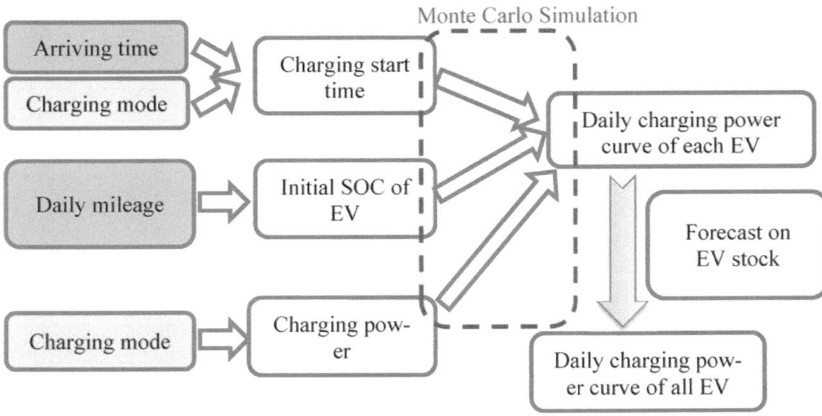

Figure 4-3 Schematic of EV charging power demand simulation

4.2 Case study: EV charging impacts on local power grid in Beijing, China

Using the above described method, EV charging impacts on local power grid with different charging modes can be evaluated. According to previous research of authors on industry development trends of EV in China, forecasted EV stocks in 2020, 2025 and 2030 in Beijing are 467,100, 1,628,600 and 3,478,000 respectively. Original power load in 2015 in Beijing was obtained, and power elasticity coefficient method was applied to forecast power load in the future. Parameters used in the forecast are shown in Table 4-1.

Table 4-1 Parameters used in forecasting power load

		Electric vehicle inventory (ten thousand)	GDP growth rate (%)	Electricity elasticity coefficient
Beijing	2015	1.04	—	—
	2020	46.71	6	0.67
	2025	162.86	5.5	0.63
	2030	347.80	5	0.6

Impacts of EV charging on local power grid in Beijing in 2020, 2025 and 2030 with any-time charging mode are shown in Figure 4-4. 6:00 am in the current day to 6:00 am in the next day (corresponding to x-coordinate 30 in the figure) is taken as the time range for the analysis.

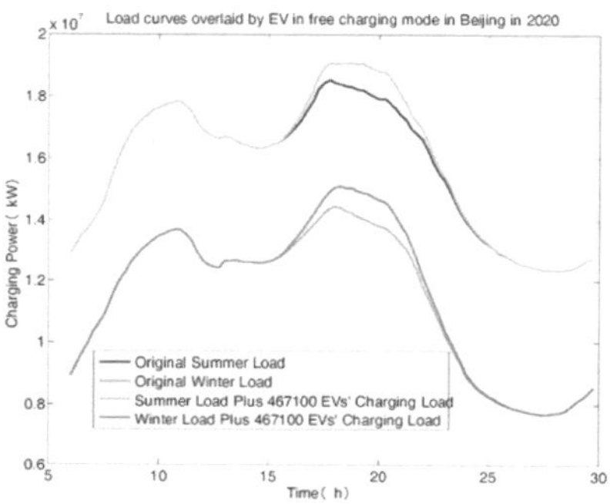

(a)

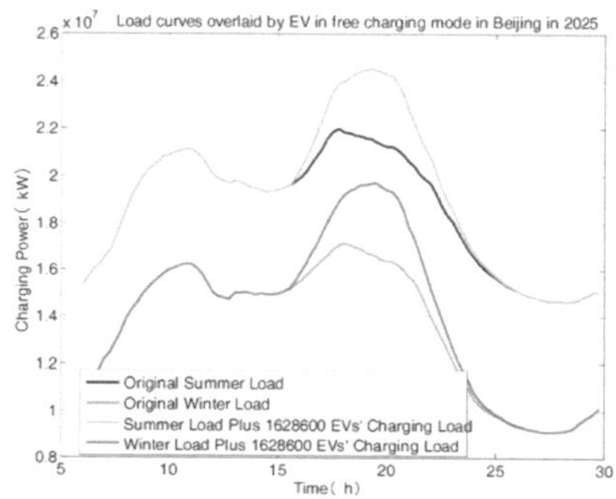

(b)

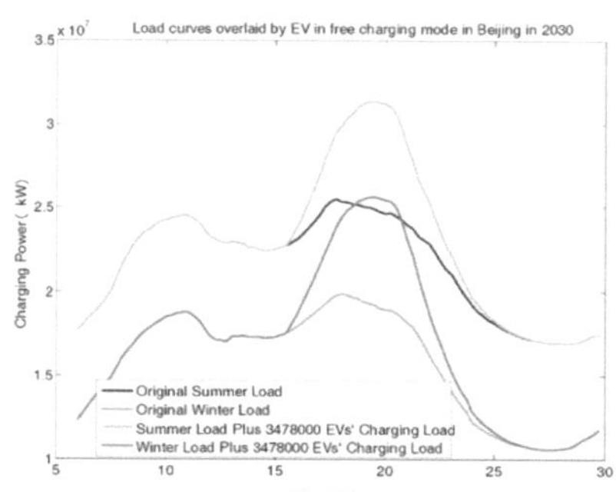

(c)

Figure 4-4 Impact of electric vehicle charging power on the power grid of Beijing with any-time charging mode in 2020, 2025 and 2030

From the simulation results it can be seen: by 2020, the increasing number of electric vehicles in Beijing with any-time charging mode would have a significant impact on the power load (since the time node before 2020 is not calculated, the situation could happen before 2020). This will lead to a higher peak load and further increase in peak-valley difference, and cases will be even worse in 2025 and 2030. It is unfavourable to the safe and economical operation of the power grid and does not support the sustainable development of electric vehicles. Therefore, it is necessary to improve the electric vehicle charging mode and make full use of the existing capacity of power grid. Figure 4-5 showed the impacts of EV charging on the power grid of Beijing with time-delayed charging and orderly charging modes in 2030.

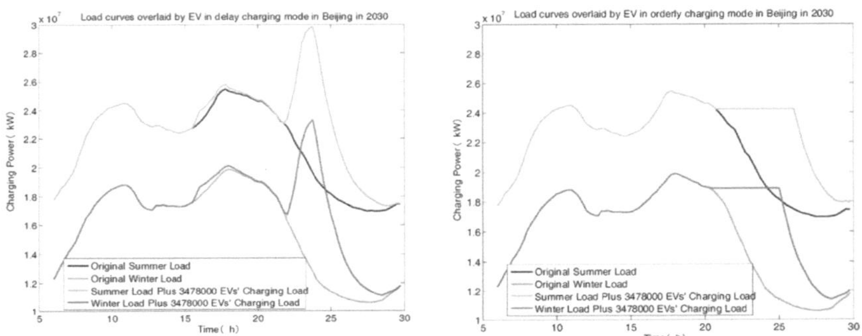

(a) time-delayed charging mode

(b) orderly charging mode

Figure 4-5 Impacts of EV charging on power grid of Beijing with different charging modes in 2030

The simulation results in Figure 4-5 indicate that, although time-delayed charging mode avoids the overlapping of charging load peaks and original load peaks, a new significantly higher load peak appears due to the great number of EV arriving in a narrow time window. Furthermore, it can be seen that the total load is still low during 1:00am – 6:00am. That implicates that the charging power demand of a certain amount of EV can be met within this time frame, thereby avoiding new load peaks and also reducing the peak-valley gaps. For cities with small amount of EV and large peak-valley gaps, time-delayed charging can be applied to avoid a new load peak.

Orderly charging overcomes the shortage of time-delayed charging. It allows as many EVs as possible to be charged with the acceptable capacity of power grid (in this paper, it is assumed that EV charging is allowed when the current load of power grid is not greater than 95% of its peak load and it should also ensure that the total load during the charging is not higher than 95% of its peak load). This mode ensures the safe operation of the power grid as well as charging demand of EV.

5 Conclusions

Currently in China NEV especially BEV has gained comprehensive attention and more and more support from the government, vehicle manufacturers and citizens. Many policies and subsidies have been released by central and local governments to promote the development of EV. However, the negative impacts of EV charging on power grid has not received sufficient recognition mainly due to the fact that not significant amounts of EV are distributed in the cities at present.

EV charging impacts on power grid with any-time charging mode, time-delayed charging mode and orderly charging mode in Beijing were studied in this paper. The any-time charging mode guarantees a maximum convenience for the user, however new higher load peaks will impact the grid. Time-delayed charging can shift charging peaks to the original load valley, and may avoid a higher load peak if the number of EV is not too large. Orderly charging can completely avoid a new higher load peak with the help of bi-directional communication between charging points and power grid even if the number of EV is huge. It is capable to satisfy the EV charging demand to the greatest extent with the acceptable capacity of the power grid. Considering the automation level of power distribution network at present in China, time-delayed charging is hopefully to be popularized, while orderly charging can be expected with the development of smart grid systems in the future.

Reference

SCPRC, 2012. Energy saving and new energy automobile industry development plan, p2-p4, State Council of the People's Republic of China, Beijing, 2012.

BMCDR, 2015[1]. Beijing administrative measures of public charging infrastructure investment and construction (Trial), p2-5, Beijing Municipal Commission of Development and Reform, Beijing, 2015.

BMCDR, 2015[2]. Notice on collecting charging service fee in Beijing, p1, Beijing Municipal Commission of Development and Reform, Beijing, 2015.

China Automotive Energy Outlook 2012, 2012.China Automotive Energy Research Center, Tsinghua University, Science Press, p225-230:2012, Beijing

Ddc.Net.cn, 2015-3-13, http://news.ddc.net.cn/newsview_58700.html

GOSC, 2015. Guiding opinions on accelerating construction of NEV charging infrastructure, p1-3, General Office of the State Council of the People's Republic of China, Beijing, 2015.

MIIT, 2009[1]. Admittance regulation for new energy vehicle manufactures and productions, p1-p3, Ministry of Industry and Information Technology of the People's Republic China, Beijing, 2009.

MIIT, 2009[2]. Auto industry adjustment and revitalization Plan, p1-p3, Ministry of Industry and Information Technology of the People's Republic China, Beijing, 2009.

MOF, et al., 2013. Notice on continued development of new energy vehicles, p1- p2, Ministry of Finance of the People's Republic of China, Ministry of Science and Technology of the People's Republic of China, Ministry of Industry and Information Technology of the People's Republic of China, National Development and Reform Commission, Beijing, 2013.

MOF et al., 2015. Notice on the financial support policy for new energy vehicles development in 2016-2020, p1-p2, Ministry of Finance of the People's Republic of China, Ministry of Science and Technology of the People's Republic of China, Ministry of Industry and Information Technology of the People's Republic of China, National Development and Reform Commission, Beijing, 2015.

MOHURD, 2015. Notice on reinforcing planning and construction of NEV charging infrastructure, p1-p2, Ministry of Housing and Urban-Rural Development of the People's Republic of China, Beijing, 2015.

NDRC et al., 2015. Development guideline of NEV charging infrastructure (2015-2020), p8-p15, National Development and Reform Commission (NDRC), National Energy Administration, Ministry of Industry and Information Technology of the People's Republic of China, Ministry of Housing and Urban-Rural Development of the People's Republic of China, Beijing, 2015.

PGBM, 2010. Temporary quantity control of passenger vehicles in Beijing, p1-p2, Beijing Municipal People's Government, Beijing, 2010.

PGSM, 2014. Regulations of quota auction for non-operating passenger vehicles in Shanghai, p1-p2, Shanghai Municipal People's Government, Shanghai, 2014.

SMPG, 2016. Temporary administrative measures of purchasing and utilizing new energy vehicles in Shanghai (2016 revision), p1-p2, Shanghai Municipal People's Government, Shanghai, 2016.

www.caam.com.cn, 2016. China new energy vehicle marketing report, 2016-04-08, http://www.caam.org.cn/hangye/20160418/1605189773.html

www.caam.com.cn, 2015. New energy vehicle ownership is increased by 170% in 2015, 2016-01-26, http://www.caam.org.cn/hangye/20160126/0905184583.html

www.chinadaily.com.cn, 2014. New energy vehicles 'turning the corner', 2014-01-11, http://www.chinadaily.com.cn/business/motoring/2014-01/11/content_17229981.htm

www.chinaequip.gov.cn, 2015. National network of major technical equipment, 2015-08-06, http://www.chinaequip.gov.cn/2015-08/06/c_134485973.htm

Yao Qiang, Yu Yongda, Zhu Guiping et al., 2015. China Resource Energy and Sustainable Development, Science and Technology Press, p432: Beijing, 2015

Zhu Guiping, Xing Yuhui, 2013. Electric Vehicle Charging Strategy in Metropolises in China. IEEE Power Energy Society General Meeting 2013, July 21-25, Vancouver, Canada.

Wie Fahrzeugbatterien die Energiemärkte revolutionieren

(Unlock the value of electric vehicles' batteries)

Marcus Fendt, The Mobility House GmbH

Abstract

Die Energiewende vorantreiben, eine emissionsfreie Zukunft ermöglichen – das ist das Ziel von The Mobility House. Durch neuartige Lade- und Energiespeicher-Lösungen verhelfen wir der Elektromobilität zum Durchbruch und reduzieren die Kosten von Elektrofahrzeugen. Unsere Technologien ermöglichen die Integration von Elektrofahrzeugen in das Stromnetz als aggregierter Schwarmspeicher fahrender Elektroautos (V2G) oder als stationäre Speicher aus Fahrzeugbatterien.

Smart charging in daily routine – expectations, experiences, and preferences of potential users

Franziska Schmalfuß*, Maria Kreußlein, Claudia Mair, Susen Döbelt, Cynthia Heller, Ramona Wüstemann, Bettina Kämpfe, & Josef F. Krems

Technische Universität Chemnitz, Institute of Psychology, Cognitive and Engineering Psychology

Wilhelm-Raabe-Str. 43, 09120 Chemnitz, Germany

* Contact corresponding author:
Phone: +49 371 531 38235;
E-mail: franziska.schmalfuss@psychologie.tu-chemnitz.de

Introduction

In the mobility sector, battery electric vehicles (BEVs) provide a promising solution to rising CO2 emissions and, in the long term, the dependence on oil. However, in the energy sector, conventionally charged BEVs provide an additional challenge as they potentially lead to higher energy consumption and a higher need for balancing energy supply and demand. In contrast, when using smart charging technologies, BEVs could become a valuable energy balancing tool in smart grids by using them as "controllable" energy consumer and energy storage. Until now, user-centered research on the acceptance of smart charging systems is rare (e.g., Pettersson, 2013; Schmalfuß et al., 2015). However, as smart charging concepts have an impact on a very essential aspect of the daily routine – mobility – a user-centered approach (Norman & Draper, 1986) when developing smart grid applications (e.g., smart charging systems) is highly recommended (e.g., Verbong, Beemsterboer & Sengers, 2013). Investigating expectations, experiences and preferences of potential users of a smart charging system and integrating the results in the developmental process, was one part of the project "Gesteuertes Laden V3.0" which was funded by the German Federal Ministry for the Environment, Nature Conservation, Building and Nuclear Safety. In this contribution, we summarize the main user studies and their results.

User research on a smart charging system

Different user studies were set up in cooperation with BMW Group, Vattenfall Europe, EWE AG, Clean Energy Sourcing AG, Fraunhofer Advanced System Technology AST and Technische Universität Ilmenau within the research project "Gesteuertes Laden V3.0". In focus groups, online surveys and a 1-year-field study including three study periods, qualitative and quantitative data were collected on preferences, expectations, attitudes regarding smart charging, acceptance, charging behaviour, experiences and future requirements. The developed and in the field trial implemented smart charging system aimed to maximize the potential of BEVs for secondary reserve power. The implementation built the valuation basis for the different studies. In the following, methods and results of selected studies are summarized and presented. For further results of the user research within the project see Döbelt, Kämpfe, & Krems (2014) or Schmalfuß et al. (2015)

Focus group

To gather first requirements on a smart charging system, a focus group was conducted in Berlin at the beginning of the project. All seven focus group participants (5 men, 2 women) had at least half a year of experience with BEVs and were therefore well informed about the daily routine with a BEV and its charging process.

Within the focus group, the general concept of a smart charging system and its consequences for society, users' charging behaviour, BEVs, etc. as well as predictability of charging events, potential participation and incentives were discussed.

In general, participants stated to be willing to use a smart charging system and that most charging events are predictable in terms of departure time, although an exact predication of departure times is challenging and there should be some sort of flexibility. This reflects one of the mentioned requirements on such a system. They also emphasized the need for a cost-benefit-balance when using smart charging, wished for few restrictions on movement freedom and flexibility for departure times. Further, they argued in favour of a mandatory option to start an immediate 'uncontrolled' charging process as well as a minimum range as 'safety buffer' that is charged unconditionally as fast as possible for emergency situations. They expressed the need for economic benefits as addition for ecological benefits that might be gained when using smart charging and a positive outcome of cost-benefit analyses. With regard to human-machine-interfaces of a smart charging system, they preferred a smartphone application.

In a next step, results of the focus group should be validated with bigger samples. Ideas regarding a smart charging system, that were derived from the focus group and further developed, should be investigated within two online studies.

Online studies

In two online surveys, preferences were investigated using questionnaires and conjoint-based measurement. Results should be used to build a reward system for participation on the smart charging system.

Respondents of the *first online study* ($N = 217$, 76% men, 24% women) were on average 36 years old and approx. 66% had experiences with controlled charging. Monetary incentives such as energy cost savings turned out to be the most attractive rewards compared to environmental-related incentives (e.g., visual feedback on the effect for the environment) or BEV-related incentives (e.g., points for earning a new battery, kilometre for a renting car). For the identification of central attributes of the reward system, respondents got a choice-based conjoint task in which they chose between four alternatives ten times. Three alternatives were different opportunities for charging varying in connection time per week (20, 40, 60 vs. 80 hours), flexibility in departure time (+/- 10, 20, 30 vs. 40 min), weekly time effort for adjusting settings (+/- 0, 1, 5 vs. 10 min) and financial incentive (2.50, 5, 7.50 vs. 10 €) and additionally a 'none' option existed if they did not favour one of the three detailed charging options. Results of the choice-based conjoint analyses showed that in comparison the connection time of the BEV with the grid and the financial benefit when using smart charging are of higher importance for the incentive scheme than the effort of adjusting settings and drivers' flexibility of the departure time.

In the *second online study* ($N = 62$, 97% men, 3% women, mean age = 41 years), the chosen attributes for the implementation of the reward system were investigated regarding their relative importance to potential users. In detail, connection time, flexibility of departure time, range buffer and minimum range when unplugging the BEV determined the potential benefit of using smart charging. Most of the respondents (92%) had experienced a BEV and about one quarter had experienced some kind of controlled charging. In 8 trials, respondents chose between four different alternatives that varied in connection time (6, 8, 10 vs. 12 hours), flexibility of departure time (+/- 5, 10, 20 vs. 30 min), range buffer (46, 53, 60 vs. 67 km) and minimum range (75, 90, 105 vs. 120 km) when unplugging the BEV and a 'none' option. Results expressed no differences in relevance of the attributes. Therefore, based on this, a weighting of attributes within the reward system was not considered. Additionally, respondents were asked for their flexibility needs/settings (in min) regarding departure time and typical charging timeframes on weekdays (Mon-Fri) and weekends. During the weekend, respondents stated higher flexibility needs compared to weekdays with relatively small (0-10 min) or medium (20 min) flexibility settings that were required by 23% and 27% respectively of the respondents (Figure 1). Most respondents (94%) named a typical charging timeframe during weekdays (on average: from 6:42 pm to 7:04 am, 9.5 hours). At weekends, 71% of the respondents stated to be able to regular charge a BEV (on average: from 15:55 pm to 8:54 am, 14.6 hours), which implies that the potential for smart charging is bigger at weekends as connection times are much longer.

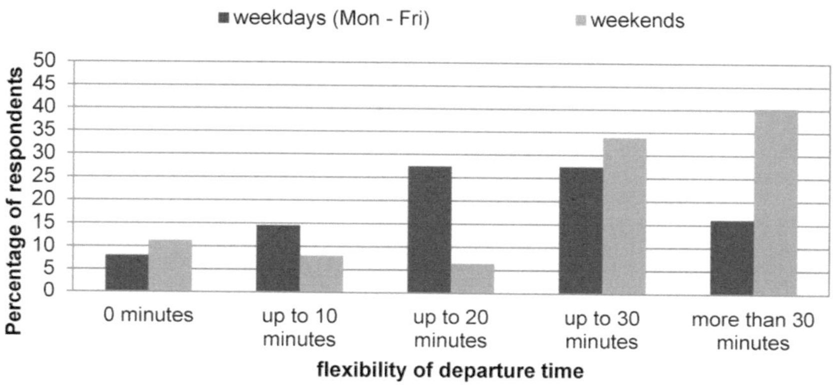

Figure 1. Percentage of respondents needing different levels of flexibility on departure times on weekdays and weekends.

Field study

The developed prototype of the smart charging system was tested extensively by BEV drivers in a 1-year field study. In the following, a summary on the implemented methods is given and selected results are presented.

Methods

Concept of the field study and the implemented smart charging system

In each of the three study periods within the 1-year field trial, 10 participants tested the smart charging system. The BEVs (BMW ActiveE was the test vehicle in the field trial) were equipped with smart charging technology. Users could adjust settings which determined the charging process via a smartphone application. In the first two study periods, web enabled smart charging boxes were installed at each participant's garage or carport[1]. In the third period, the charging boxes were installed at participant's workplaces, so that they mainly could charge during daytime. In the present contribution, the first two study periods are in the focus of the report.

1 In the project, the smart charging technology was implemented in an ActiveE and a home charging station by BMW Group and Vattenfall Europe. This technology could be transferred to other types of BEVs and charging stations as long as there is an internet connection available at the charging location. The smart charging stations that were used also needed a water-proof installation location, but this precondition could be eliminated by investing in the production of the charging boxes.

The field studies consisted of three phases: 1) a habituation phase, in which participants should first get used to the BEV and normal charging Figure 1), followed by 2) a baseline with uncontrolled charging option and 3) a period with controlled charging (CC) only. Data assessment via interviews and questionnaires took place before activating 'controlled charging' (CC) and after using the smart charging system for several weeks. The timeline of the field study is displayed in Figure 2.

In the first data assessment, an online questionnaire (LimeSurvey 2.00 +) was answered by the participants. At all other assessment points, semi-structured interviews and paper-pencil questionnaires were used to collect data, for instance, on expectations, experiences, acceptance, suitability for daily life, trust and motivation as well as charging behaviour. Besides, charging and event diaries were filled out (Figure 2).

Shortly after the first data assessment, participants got the option to switch between two charging modes at any time: conventional uncontrolled charging (UC; i.e. charging starts immediately after plug-in) and controlled charging (prototype system). The implemented CC condition was programmed to charge the BEVs whenever the energy demand in the grid is low (e.g., late night hours) and the need for regulation power is high. Calculated charging schedules were fitted to a simulated grid load (i.e., based on historic data, a grid was simulated simultaneously) and integrated with a planned departure time, the flexibility for users' departure time[2], the minimum state of charge (MinSOC) that must be guaranteed at departure time ($50\% \leq$ MinSOC $\leq 100\%$) and a minimum state of charge that must be reached as fast as possible (Safety Buffer, $30\% \leq$ Safety Buffer $\leq 45\%$). Users could adjust all settings via a smartphone application. In addition to functioning as the smart charger interface, the application served as remote access to the vehicle, providing information such as state of charge, range, charging process on/off, as well as energy consumed and costs for all previous charging events. Before mentioned studies (e.g., focus group) as well as other projects (e.g., Pettersson, 2013) showed a user preference for a mobile application as user interface.

2 To meet the result of the above described focus group the function of setting the flexibility of departure time was implemented in the system.

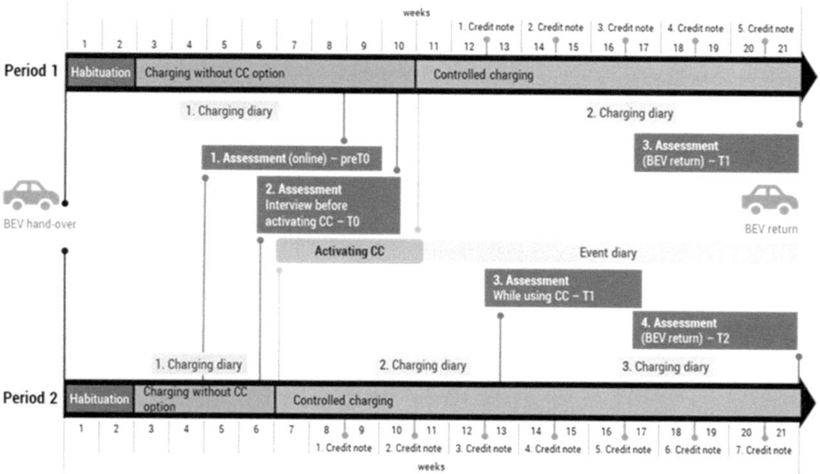

Figure 2. Timelines of the two study periods.

Reward mechanisms using points were used to achieve users' motivation for plugging-in their BEV as long as possible, selecting rather small safety buffers and low levels of MinSOC. Participants could reach a maximum of 40 points per charging event which were displayed in the smartphone application. Points were converted into Euros every 2 weeks (1 point = 2 Euro cents). A second face-to-face interview (T1) was executed after several weeks in which participants had the opportunity to use CC. Within the CC testing phase, several technical problems occurred, such as a 3-day server crash, delayed feedback on earned points per charging event due to internet connection failures and problems with charging infrastructure. Participants returned their BEVs in the first phase at T1 and in the second phase at T2, the third point of data collection that only took place in the second phase. In the present contribution results from T0 and T1 are in the focus. At T1, participants of both phases had a comparable length of experience with CC.

Material

Demographic variables (age, gender, highest level of education, household net income, experience with vehicles with electric drive and monthly driving distance) were assessed within the online application form.

Questionnaires and semi-structured interviews were processed before and after using the smart charging system were implemented to gather participants' preexisting motivation, expectations, as well as their experience and utilization of CC before (preT0,

T0) and after using the smart charging system (T1, T2). Interviews were digitally recorded, transcribed verbatim and coded according to thematic analyses by Braun and Clarke (2006), using the software tool MAXQDA 11 (VERBI Software, 2014).

At T0 and T1, the *questionnaires* contained the following scales and items. In order to assess motivation for using CC, participants were instructed to distribute a total of 100 points to 3 given categories ("contribution to grid stability", "contribution to integration of renewable energies" and "reward for participation") and one open-ended category. The 12-item unidimensional scale by Jian, Bisantz and Drury (2000) was utilized for the assessment of trust in automated systems –the smart charging system – using a seven point Likert scale ranging from 1 (*fully disagree*) to 7 (*fully agree*). An overall trust score was obtained by calculating the mean of all 12 items. Furthermore, participants rated perceived effectiveness. The statement "the implementation of CC is acceptable" assessed the acceptability of CC. Additionally, willingness to use CC in the project and afterwards was assessed in the questionnaire. Different statements perceived benefits (e.g., financial savings) and costs (e.g., less flexibility) were also included for evaluation (see Schmalfuß et al., 2015). Furthermore, participants rated the attractiveness of different forms of incentives (see Figure 3) on a 10-point scale (1 = *very unattractive* to 10 = *very attractive*). All scales mentioned above were included in the questionnaires at T0 and T1. If not stated differently, a 6-point Likert scale ranging from 1 (*completely disagree*) to 6 (*completely agree*) was utilized.

The questionnaire also included an open-ended question about the average number of charging events the participant initiated each week. Only at T1, the number of controlled as well as uncontrolled charging events was assessed and participants rated different reasons for not using CC (e.g., technical problems).

Within the semi-structured *interview* at T1, participants were asked "Please summarize your experiences with CC" and "Do you have any suggestions for improvement?" after experiencing CC.

Charging diaries (following Krems, 2011 and Franke & Krems, 2013) were implemented in order to assess charging frequencies, location, mode and reasons for charging behaviour within a two-week period (see Figure 2). *Event diary* (following Schmehl et al., 2011) gave participants' the opportunity to note special events, positive and negative, relating to the smart charging system (see Figure 2).

Participants

More than 300 potential users applied for participating in the project via an online tool. The field study was advertised via different media channels (internet, local newspaper, local radio). Twenty participants (18 men, 1 woman) were selected that satisfied various criteria (e.g., have a water-proof location with internet connection on which a smart charging box can be installed, willingness to pay a monthly leasing rate) and showed a wide variety in sociodemographic variables. They were on average 49 years old (*Min* = 32, *Max* = 61) and the majority of the participants were highly educated (14 hold a university degree, 3 are master craftsmen). Two households used only the BEV during the field study. All others had one ($n = 8$) or more additional vehicles ($n = 12$). BEV drivers in other studies showed similar distribution on sociodemographic variables (e.g., Hjorthol, 2013); participants are mainly middle-aged males, highly educated and had the opportunity to charge at home. Therefore, the sample was assumed to be representative of typical private BEV buyers, who have the opportunity to use a smart charging system at home.

Results

Results of the *interview* showed that users made positive as well as negative experiences with the smart charging system. On the positive side ($N = 14$), for instance, they stated that the system was easy to use. On the negative side, users described problems while charging (e.g., server availability, $N = 8$), negative experiences with the application and the whole system in general (e.g., system needs further development, $N = 4$) and reduced flexibility of their mobility ($N = 3$) as well as negative experiences with the reward system (e.g., low financial savings, $N = 3$). Two users stated that controlled charging did not match with their personal needs.

Questionnaire results showed that on average users were positive about the system (Table 1); they evaluated it as trustworthy, (rather) acceptable and efficient. This general evaluation of the smart charging system was reflected in users' willingness to use controlled charging within the project and beyond. Except for suitability for daily life which decreased from T0 to T1, none of these variables changed significantly over time during the field study.

Test users were convinced that they are more environmental friendly when using smart charging and that they contribute to 'green energy supply' and 'grid stability'. After testing smart charging, they were more skeptical regarding 'a cheaper energy supply' for charging their BEV and 'lower financial costs'. At both points of data assessment, users neither agreed nor disagreed that planning charging processes is a slightly greater challenge when using controlled charging compared to uncontrolled charging. At the same time, users stated to be rather equally flexible independently of the charging mode. Test users (rather) disagreed that their 'freedom of movement is restricted compared to others' and that they make less spontaneous trips. After testing controlled charging, participants showed higher appraisal to the statement that the state of charge is on average lower when using smart charging.

At both points of data collection, participants indicated that they were mainly motivated to use controlled charging because of the opportunity to 'contribute to the integration of renewable energy into the grid' (M_{T0} = 43.9, M_{T1} = 47.3). On second place was the statement 'I contribute to the stability of the grid' (M_{T0} = 33.3, M_{T1} = 25.8). The statement, "I get a reward for participation", received the third most points (M_{T0} = 22.9, M_{T1} = 25.8). Only at T1, points were given to other options, such as "taking part in a field study".

Table 1. Questionnaire results from T0 and T1.

Point of data collection Assessed concepts	Mean scores T0	 T1
Perceived benefits		
... I have lower financial costs than non-participants [a]	5.0	4.0
... I contribute to grid stability [a]	4.9	4.8
... I have a cheaper energy supply [a]	4.5	3.8
... my EV is more environmentally friendly [a]	5.1	5.0
... I contribute to the green energy supply [a]	4.7	4.7
Perceived costs		
... my freedom of movement would be unfairly restricted compared to others. [a *]	2.5	2.4
... it is a big challenge to plan car usage (charging events, charging durations). [b]	3.7	3.7
... the state of charge of my ActiveE is lower on average [b]	3.9	4.5
... I make fewer spontaneous trips [b]	3.5	3.2
... I have higher effort to plan trips [a]	3.8	3.6
... I am as flexible as before (i) [b]	2.1	2.3
... I have to charge more often in between [b]	3.0	3.4
... I have higher effort to plan charging events compared to conventional charging [a *]	4.3	4.1
Trust	5.6	5.4
General Evaluation		
Suitability for daily life	4.4	3.9
Effectiveness	4.9	4.5
Acceptability	4.6	4.2
Willingness to use CC in the project	5.6	5.4
Willingness to use CC beyond the project	5.0	5.3

Note. (i) = inverted item;
[a] = items began with "When I use CC, compared to normal charging,…";
[b] = items began with "When I charge my ActiveE controlled…";
* items retrieved from Eckhardt et al. (2011)

As in the earlier studies within the project, financial savings were the prioritized options (e.g., cost savings on energy needed for charging) as reward for using controlled charging compared to, for instance, ecological benefits (Figure 3).

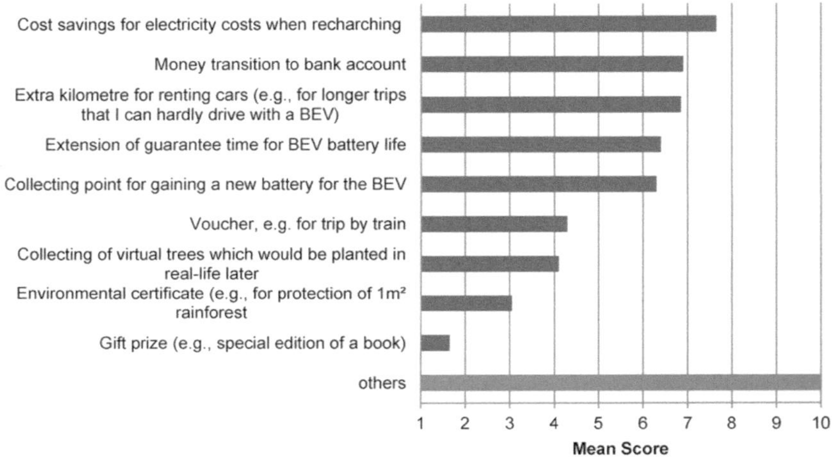

Figure 3. Different types of incentives and their evaluations

For most charging processes participants stated to have used smart charging: On average they started 6.2 charging events per week and chose for 5.3 of them 'controlled charging' instead of 'normal' charging. Results of *charging diaries* are in line with this (5.4-6.9 charging event per week and 3.7-5.9 with 'controlled charging' mode) and showed that on average users plugged in for 11 to 12.5 hours and charged mainly at home (81%-99% of all charging events within the two weeks of filling in charging diaries).

Participants' answers regarding future requirements showed that the system needs further improvement in server/connection stability ($N = 9$) and therefore reliability. Charging duration should be reduced ($N = 6$) and the flexibility of rewarded charging times should be enhanced ($N = 5$). The reward system was another aspect mentioned for improvements; higher profitability was requested ($N = 4$). In the opinion of some users, future systems might also integrate innovative technologies such as vehicle-to-grid ($N = 3$) or inductive charging ($N = 1$). Additionally, push-up notifications in cases of system failures etc. ($N = 6$) and an additional user access (e.g., second device, $N = 2$) to interact with the system was desired by several users.

Summary and Implications

The findings of the different studies highlight the importance of a user-centred approach and were used to further develop and improve the system idea as well as its prototype. On the basis of different user studies, for instance, the application was (further) developed as well as the reward system planned and then realized. The whole smart charging system was intensively tested in real life and users could gain experience over several months. Results of the field study show that a smart charging system is useable and acceptable in daily routine, but improvements are needed. Valuable indications in order to increase consumer acceptance of smart charging systems in future ecomobility systems were gained. Future systems need to be further developed, especially in terms of reliability. In addition, shorter charging durations or a fast charging option should be realized in order to address the users' requirements of more flexibility of charging processes and freedom of movement. Further potential for improvement was found for the reward system (i.e., higher profitability) and the application (i.e. more information about BEV and state of charge). Technical improvements, such as the integration of vehicle-to-grid or providing a second interface (e.g., additional smartphone login, online portal) might have a positive effect on the user experience and suitability for daily life.

Potential users as well as test users prioritized monetary incentives. In the present implementation, a reward system including one type of monetary incentives was tested and appeared to motivate users. The effectiveness of the developed reward system, however, needs further investigation and comparisons with other options should be drawn. Test users were also critical about the reward system and requested a higher profitability which is hardly possible from the viewpoint of market economy. Therefore, one solution is to develop even 'smarter' charging systems with learning mechanisms, so that the user involvement needed is lower and the cost-benefit analysis results in a more positive evaluation of the system. It should be investigated if they might be the more economic systems from the perspective of market economy.

The project with its developed and implemented prototype was a great start in addressing grid stability problems by utilizing BEVs and controlled charging. Now, a large-scale solution needs to be developed and introduced to the market.

Acknowledgements

This study is part of the research project "Gesteuertes Laden V3.0" which was funded by the German Federal Ministry for the Environment, Nature Conservation, Building and Nuclear Safety. We thank our consortium partners BMW Group, EWE AG, Clean Energy Sourcing AG, Fraunhofer Advanced System Technology AST, Technische Universität Ilmenau and the associated partner Vattenfall Europe that realized the smart charging system and supported the study.

References

Braun, V., & Clarke, V. (2006). Using thematic analysis in psychology. *Qualitative Research in Psychology 3*(2), 77-101. doi:10.1191/1478088706qp063oa.

Döbelt, S., Kämpfe, B. & Krems, J. F. (2014). Smart Grid, Smart Charging, Smart Privacy? An Empirical Investigation of Consumers' Willingness to Provide Smart Charging Information, *Tagungsband ComForEn 2014*, 29-37.

Vattenfall Europe Innovation GmbH (2011). Verbundprojekt: Steigerung der Effektivität und Effizienz der Applikationen Wind-to-Vehicle (W2V) sowie Vehicle-to-Grid (V2G) inklusive Ladeinfrastruktur (Gesteuertes Laden V2.0). Hamburg,. Retrieved from: https://www.tib.eu/en/search/id/TIBKAT%3A68125291X/Gesteuertes-Laden-V2-0-FuE-Programm-F%C3%B6rderung-von/ (24.02.2016).

Franke, T., & Krems, J. F. (2013). Understanding charging behaviour of electric vehicle users. *Transportation Research Part F Traffic Psychology and Behaviour, 21*, 75-89. http://dx.doi.org/10.1016/j.trf.2013.09.002

Hjorthol, R. (2013). *Attitudes, Ownership and Use of Electric Vehicles – A Review of Literature.* TØI, Institute of Transport Economics, Norwegian Center for Transport Research. Report 1261/201.

Jian, J. J., Bisantz, A. M., & Drury, C. G. (2000). Foundations for an empirical determined scale of trust in automated systems. *International Journal of Cognitive Ergonomics, 4*(1), 53-71. doi:10.1207/S15327566IJCE0401_04.

Krems, J. F., et al. (2011). MINI E powered by Vattenfall V2.0. Schlussbericht zum Forschungsvorhaben Verbundprojekt: MINI E powered by Vattenfall V2.0 (Abschlussbericht, Förderkennzeichen 16EM0070). Chemnitz : Technische Universität Chemnitz.

Norman, D. A., & Draper, S. W. (1986). *User-centered system design: New perspectives in human-computer interaction.* Hillsdale, NJ: Lawrence Erlbaum Associates.

Pettersson, S. (2013). *ELVIIS – Final Report to Göteborg Energi Forskningsstiftelse,* ICT Sweden Viktoria https://www.viktoria.se/sites/default/files/pub/www.viktoria.se/upload/publications/elviis_final.pdf.

Schmalfuß, F., Mair, C., Döbelt, S., Kämpfe, B., Wüstemann, R., Krems, J.F., & Keinath, A. (2015). User responses to a smart charging system in Germany: Battery electric vehicle driver motivation, attitudes and acceptance, *Energy Research & Social Science, 9*, 60-71. doi:10.1016/j.erss.2015.08.019.

Schmehl, S., et al. (2011). Directed cultural probes: Detecting barriers in the usage of public transportation. [Buchverf.] P. Campos, et al. *Human-Computer Interaction – INTERACT 2011.* Heidelberg, Berlin : Springer.

VERBI Software (2014). MAXQDA: Reference Manual for the Text Analysis Software, Marburg, 2014.

Verbong, G. P., Beemsterboer, S., & Sengers, F. (2013). Smart grids or smart users? Involving users in developing a low carbon electricity economy. *Energy Policy,* 117-125.

TRAFFIC AND ENERGY SYSTEMS

Energie- und Verkehrssystem wachsen zusammen

(Energy and transport systems grow together)

Ulrich Wagner

Lehrstuhl für Energiewirtschaft und Anwendungstechnik, TU München
Forschungsstelle für Energiewirtschaft e.V., München

Inhalt

Die Stromversorgung für Elektromobilität ist mit einem Verbrauch von weniger als 2 TWh pro Million Fahrzeugen ist auch auf längere Sicht kein Mengenproblem, sondern lässt vor allem Leistungsengpässe auf der Verteilnetzebene erwarten. Daher ist es zunehmend wichtig, Elektrofahrzeuge als flexible Verbraucher mit erzeugungsgerechten Ladestrategien und ggf. auch als Spitzenleistungsspeicher (vehicle to grid) einzuplanen; so können sie zukünftig einen wesentlichen Beitrag für die verstärkte Nutzung und die Netzintegration erneuerbarer Energien leisten.

Der Beitrag behandelt sowohl die energiewirtschaftlichen Rahmenbedingungen für Elektrofahrzeuge als auch die mobilitätsrelevanten Rahmenbedingungen für das Elektrizitätssystem. Daraus lässt sich ableiten, an welchen Stellen sich beide Systeme aufeinander zu entwickeln müssen und wie sie von einer stärkeren Kopplung profitieren können.

Heutige und zukünftige Rahmenbedingungen:

Das Elektrizitätssystem in Deutschland hat in seiner 150-jährigen Geschichte einen weltweit hohen Stand bzgl. Effizienz und Versorgungssicherheit erreicht. Aufgebaut wurde dieses System zu Zeiten monopolistischer Strukturen der Stromerzeugung, die ganz überwiegend auf großen zentralen Erzeugungseinheiten in der Nähe der großen Verbrauchszentren basierten. Mit der Liberalisierung des europäischen Strommarktes im Jahr 1998, den internationalen Klimaschutzzielen nach Kyoto und weiteren Zielen und Maßnahmen aus dem deutschen Energiekonzept muss in den nächsten Jahrzehnten eine deutliche Anpassung bzw. Neustrukturierung des Stromsystems erfolgen. Das betrifft die Integration der erneuerbaren Energien aus dezentraler Erzeugung, aber insbesondere auch große zentrale Anlagen wie Wind-Offshore-Parks. Diese neue Erzeugungsstruktur erfordert u.a. eine Anpassung des Stromnetzes auf allen Spannungsebenen, ganz besonders auf der Verteilnetzebene mit seiner zukünftig besonders großen Vielfalt an dezentralen Stromerzeugern (PV, BHKW, etc.) und zusätzlichen Verbrauchern (Elektrofahrzeuge, Wärmepumpen, etc.). Mittelfristig werden zudem Speicher zum Ausgleich der zeitlichen Asynchronitäten zwischen fluktuierender Erzeugung und Verbrauch benötigt.

In Bild 1 ist die Primärenergiebilanz für Deutschland im Jahr 2012 dargestellt. Von 100 % Primärenergie entfallen knapp 35 % auf Eigenverbrauch und Verluste im Energiesektor sowie den Nichtenergetischen Verbrauch. Die verbleibenden zwei Drittel für den gesamten Endenergiebedarf teilen sich zu je etwa gleichen Teilen auf Industrie, Haushalt und Verkehr auf, der Sektor GHD (Gewerbe Handel Dienstleistungen) liegt im Verbrauch nur etwa halb so hoch. Definiert man den Nutzungsgrad pro

Sektor als das Verhältnis aus Nutzenergie pro Endenergieeinsatz, so wird schnell deutlich, dass die stationären Sektoren Werte zwischen 60 und 70 % erreichen, während im Verkehrssektor (einschließlich Bahnen) knapp 20 % erreicht werden. Diese Zahlen machen deutlich, wie hoch angesichts der ambitionierten Klimaziele der Handlungsdruck in allen Bereichen, speziell im Mobilitätsbereich ist.

Im Verkehrsbereich dominiert seit jeher der Einsatz von Mineralölprodukten, gleichzeitig ist er heute der größte Verbraucher dieser Energieträger. Bild 2 zeigt die Entwicklung des Primärenergiebedarfs in Deutschland nach Energieträgern. Der Verbrauch ist im Zeitraum von 1990 bis heute nur marginal zurückgegangen. Die Verteilung der Primärenergiearten ist im internationalen Vergleich sehr diversifiziert, was einen grundsätzlich Vorteil für die Versorgungssicherheit darstellt. Wesentliche Anteile des Mineralölverbrauchs entfallen auf den Verkehrssektor (entsprechend ca. 28 % des Endenergiebedarfs). Elektrische Energie spielt im Mobilitätssektor bis heute eine nur unbedeutende Rolle, namentlich für den Bahnverkehr. Über die Jahre ist eine Verschiebung der Energieträgerstruktur weg von Kohle und Kernenergie hin zu Erdgas und erneuerbaren Energien zu verzeichnen. Das wirkt sich besonders stark in der Elektrizitätswirtschaft aus, wo in den kommenden Jahren die Erzeugungs-, Transport- und Speicherinfrastrukturen entsprechend umgestaltet werden müssen.

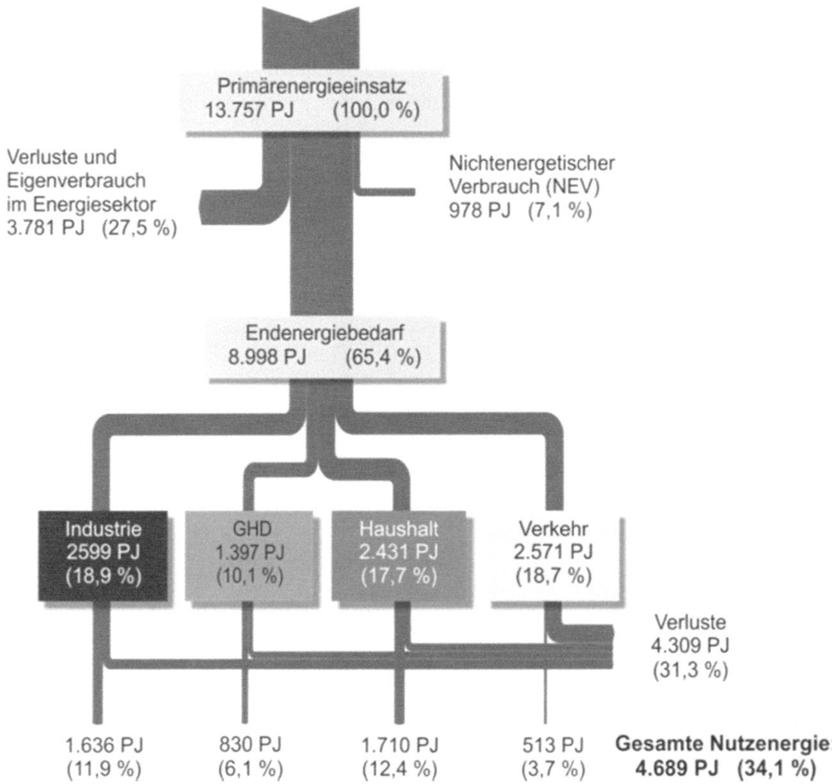

Bild 1 Energiebilanz für Deutschland im Jahr 2012

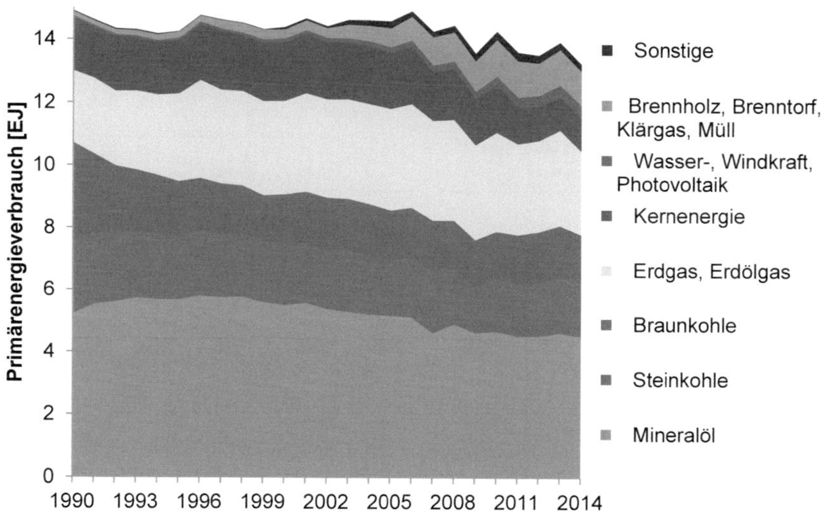

Bild 2 Entwicklung des Primärenergiebedarfs in Deutschland

Bild 3 zeigt den im Gegensatz zur Primärenergie diametral unterschiedlichen Trend bei der Entwicklung der spezifischen Verkehrsleistungen nach Verkehrsträgern in Deutschland. Die Verkehrsleistung hat sich innerhalb von 60 Jahren kontinuierlich auf heute rund 14.000 km pro Einwohner und Jahr gesteigert, rund 80 % der Verkehrsleistung entfallen auf den motorisierten Individualverkehr.

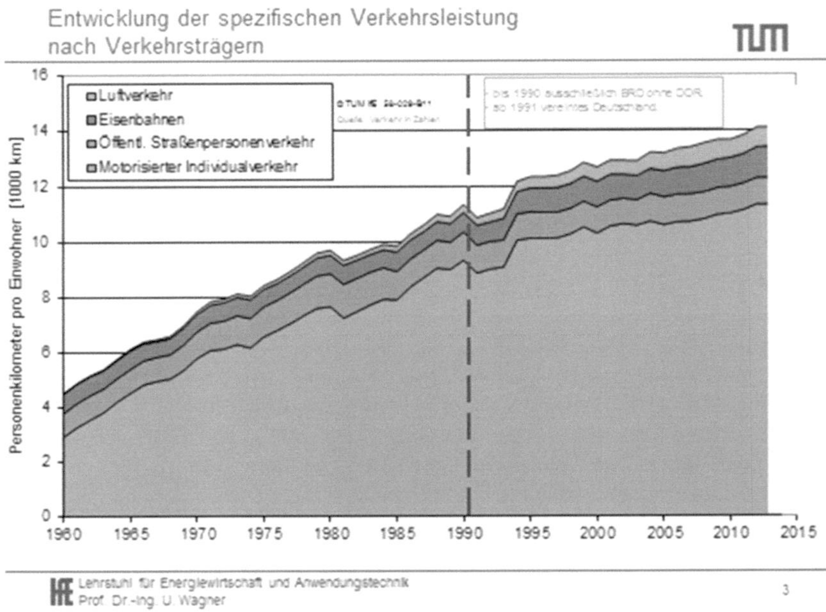

Bild 3 Entwicklung der spezifischen Verkehrsleistungen nach Verkehrsträgern

Wenn man unterstellt, dass die hoch gesteckten politischen Klima-Ziele auch im Verkehrsbereich erreicht werden müssen und die Mobilität von Personen und Gütern nicht begrenzt werden soll, sind umfassende technische und logistische Maßnahmen erforderlich. Von großer Bedeutung sind dabei auch flächendeckende Information über Bedeutung und Auswirkung des individuellen Verhaltens mit Mobilität.

Bild 4 zeigt dazu die Entwicklung des motorisierten Individualverkehrs nach Fahrzwecken. Über die Jahrzehnte entfallen konstant nur rund 40 % der Personenkilometer auf von außen vorgegebene berufliche oder Ausbildungszwecke. Rund 60 % der Verkehrsleistung dienen für Einkauf, Freizeit und Urlaub und sind damit von jedem Individuum zu jedem Zeitpunkt frei gestaltbar: darin steckt ein sehr hohes Einsparpotenzial.

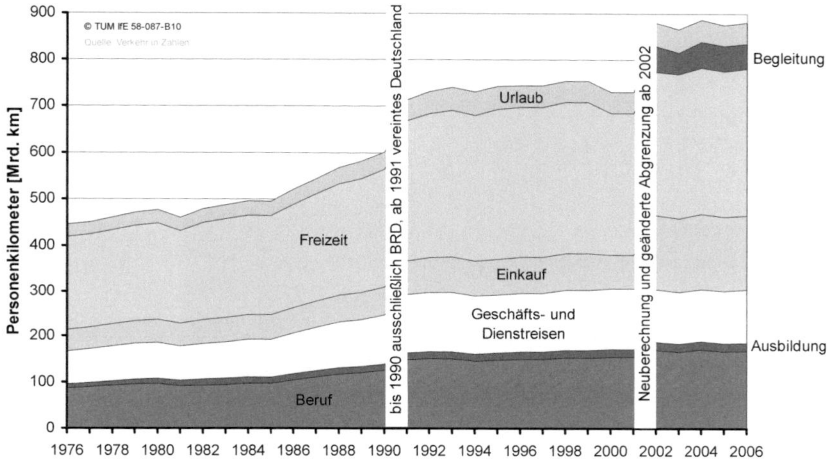

Bild 4 Entwicklung des motorisierten Individualverkehrs nach Fahrzwecken

Als Beispiel aus diesem Bereich zeigt Bild 5 die starke Abhängigkeit des persönlichen Energieverbrauchs von der Entfernung zwischen Wohnort und Arbeitsort an einem typischen Rechenbeispiel. Demnach benötigt man für eine 100 m²-Wohnung in einem Mehrfamilienhaus und 5 km Wegstrecke zur Arbeit nur rund ein Viertel der Energie für Wohnen und Fahrten als im Falle eines 150 m²-Einfamilienhauses in 50 km Entfernung vom Arbeitsort.

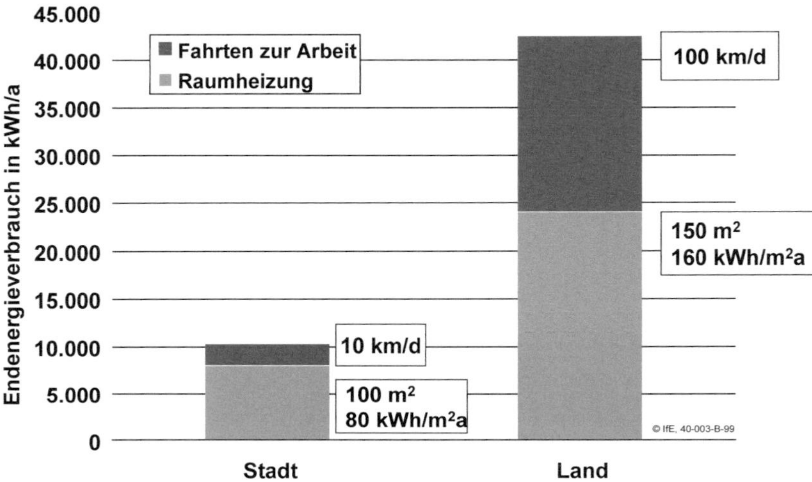

Bild 5 Energiebedarf für Wohnen und Mobilität

Diese gesellschaftlichen Kriterien sind nur schwer zu beeinflussen und zu modellieren. Daher sind Lösungen zu suchen, die möglichst unabhängig davon durch innovative Technik und Logistik (z.B. Städteplanung, Verkehrsleitsysteme, Car-Sharing) den Mobilitätsbedarf ressourcen- und klimaschonend decken.

Alternative Antriebstechniken

Im Folgenden soll ein kurzer Überblick über alternative Antriebe und deren energetische Qualität gegeben werden. In Bild 6 sind schematisch heutige und zukünftige Antriebskonzepte bzw. Kraftstoffe gezeigt. Auf der Antriebsseite (rechts im Bild) sind die beiden verfügbaren Optionen dargestellt: Antriebe mit Verbrennungsmaschinen (wandeln Moleküle in Fahrenergie um) und Elektromotoren (wandeln Elektronen um). Für bestimmte Einsatzzwecke und in der Anfangsphase der Elektromobilität ist auch die Kombination als Hybridantrieb sinnvoll. Der Tank versorgt den Antrieb mit Kraftstoff, von konventionellen Mineralölprodukten über Erdgas, Biokraftstoff, Wasserstoff oder andere synthetischen Kohlenwasserstoffen. Je nach Art und Zustand des Kraftstoffs kommen sehr unterschiedliche Ausführungen an Tanks (atmosphärisch, Hochdruck, Niedertemperatur, Anlagerung an Hydride) zum Einsatz.

Bei der Verbrennungskraftmaschine bestehen Optimierungspotentiale in einer (überschaubaren) weiteren Steigerung des Wirkungsgrades, sowie in der Verwendung von Kraftstoffen mit weniger Treibhausgaspotential wie Erdgas, Biokraftstoff, regenerativ erzeugtem Wasserstoff oder Alkoholen.

Elektrische Antriebe können direkt über einen Batteriespeicher oder von einer Brennstoffzelle mit Elektronen versorgt werden. Der Qualität des Batteriespeichers kommt dabei eine Schlüsselrolle zu; sie bestimmt die Einsatzmöglichkeiten, die Praxistauglichkeit und die Kosten des E-Fahrzeugs. Die energetische Fahrzeug-Effizienz ist hier dank hoher Nutzungsgrade aller Komponenten am höchsten. Das Brennstoffzellen-Fahrzeug erreicht zwar nicht so hohe Nutzungsgrade, dafür ist der Aktionsradius deutlich größer (H2-Infrastruktur vorausgesetzt), und es können wie beim Batteriefahrzeug erneuerbare (Überschuss-)Energien bevorzugt genutzt werden.

Setzt man die jeweils typischen energetischen Nutzungsgrade für die Bereitstellung und Nutzung der Komponenten an, so ergibt sich der in Bild 7 gezeigte spezifische Primärenergieverbrauch (Well to Wheel Analyse) für die heutige Generation von Fahrzeugen mit Diesel-, Benzin-, Batterie- oder Brennstoffzellenantrieb. Der jeweils linke Säulenteil zeigt den Bedarf an Endenergie im Fahrzeug zzgl. des Energieverbrauchs der vorgelagerten Kette (Gewinnung, Raffinerie, Stromerzeugung, Gastransport etc.). Der rechte Säulenteil zeigt die zugrundeliegenden Primärenergieträger.

Im Falle von Diesel- und Benzin-Fahrzeugen addieren sich zum Verbrauch im Fahrzeug mit 44 bzw. 48 kWh/100km etwa 10% aus der vorgelagerten Kette. Der dominierende Energieträgeranteil ist Erdöl, in der vorgelagerten Kette kommen Strom und biogene Rohstoffe zum Einsatz, insgesamt ergibt sich ein PE-Bedarf von 50 bzw. 55 kWh/100km. Das Batteriefahrzeug hat vergleichsweise einen Fahrenergieverbrauch von weniger als einem Drittel. Die gesamte Primärenergiebilanz hängt wesentlich von der Art der Strombereitstellung ab: im Bild ist als Referenzfall der aktuelle Stromerzeugungsmix zugrunde gelegt, damit steigt der PE-Verbrauch auf rund 42 kWh/100km. Dabei gilt es zu berücksichtigen, dass der Primärenergiemix schon heute über 30 % erneuerbare Energie enthält. Dieser Anteil wird sich in den nächsten Jahren weiterhin deutlich erhöhen, damit profitiert die Elektromobilität unmittelbar von den steigenden Anteilen CO_2-freier Stromerzeugung und kann als aktive Komponente zur verbesserten Integration erneuerbarer Energien im Stromsystem beitragen.

Gleiches gilt sinngemäß auch für die Erzeugung von Wasserstoff für Brennstoffzellen-Fahrzeuge (BSZ), die z.B. mittels Elektrolyse aus Windenergie und Photovoltaik oder mittels Dampfreformierung aus biogenen Rohstoffen erfolgen kann. Der Fahrenergieverbrauch von BSZ-Fahrzeugen liegt bei rund 28 kWh/100km.

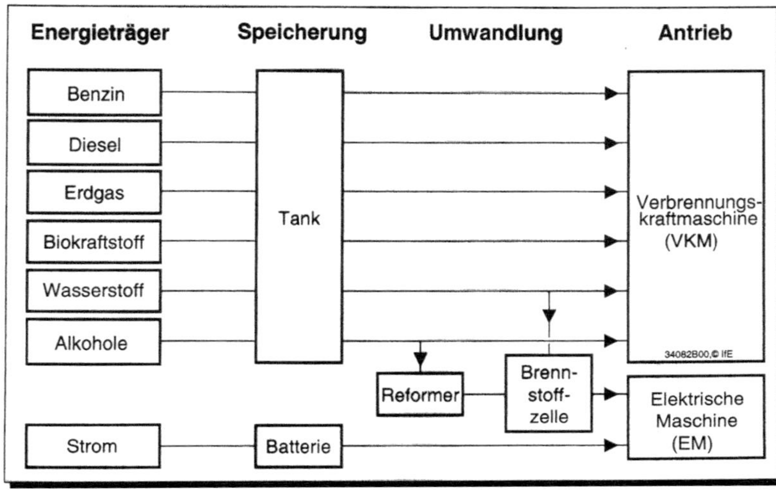

Bild 6 Überblick über verfügbare neue Antriebskonzepte

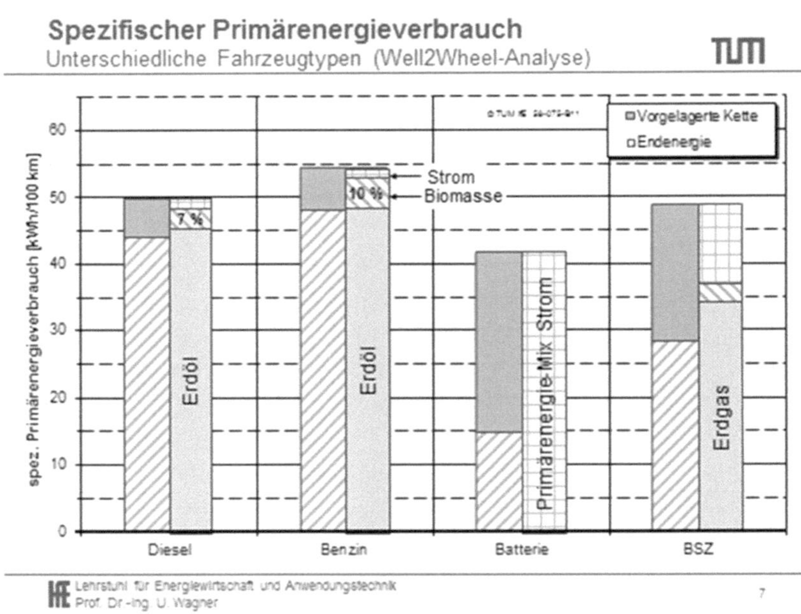

Bild 7 Energiebilanzen für verschiedene Antriebskonzepte

Zukünftig wird es starke Wechselwirkungen zwischen elektrisch betriebenen Fahrzeugen mit Batterien bzw. Brennstoffzellen und dem Energiesystem geben. Auch wenn der zusätzliche Strombedarf für eine Million Elektrofahrzeuge unter 2 TWh pro Jahr liegt, entstehen neue technische Herausforderungen auf der Verteilnetzebene. Das Elektrofahrzeug muss eine aktive und „netzdienliche" Rolle im Stromnetz übernehmen, um Ladevorgänge zeitlich flexibilisieren zu können (natürlich stets unter Einhaltung der Nutzeranforderungen an die Fahrzeugverfügbarkeit). Damit kann das Fahrzeug je nach Ladebetriebszustand positive oder negative Primär- und Sekundärregelleistung erbringen, als erweitere Option auch Energie aus der Batterie in das Netz einspeisen. Eine weitere Einsatzoption ist die Spannungsstützung in ausgedehnten Netzen, wodurch Netzausbau und Blindleistungskompensation vermindert werden können. Diese Funktionalitäten werden unter dem Begriff V2G (Vehicle to Grid) zusammengefasst.

Damit erweitert sich das Portfolio möglicher Einbindung und Geschäftsmodelle ganz erheblich. In der Studie „Merit Order Matrix 2030" /FfE 2016/ wurde das Elektrofahrzeug als funktionaler Speicher neben anderen Möglichkeiten der Energiespeicher und Lastbeeinflussung jeweils aus betriebs- und volkswirtschaftlicher Sicht bewertet. Bild 8 zeigt als zusammenfassendes Ergebnis die resultierende Merit Order Matrix in Indexdarstellung: Power to Heat und industrielles Lastmanagement sind schon heute die betriebs- und volkswirtschaftlich attraktivsten Pfade. Auch das Vehicle to Grid weist mit den Geschäftsmodellen kurzfristiger Stromhandel und Regelleistungsbereitstellung eine sehr hohe Rentabilität für das System auf, allerdings ist wegen mangelnder Infrastrukturen und fehlender Geschäftsmodelle die Wirtschaftlichkeit für den Akteur noch nicht gegeben.

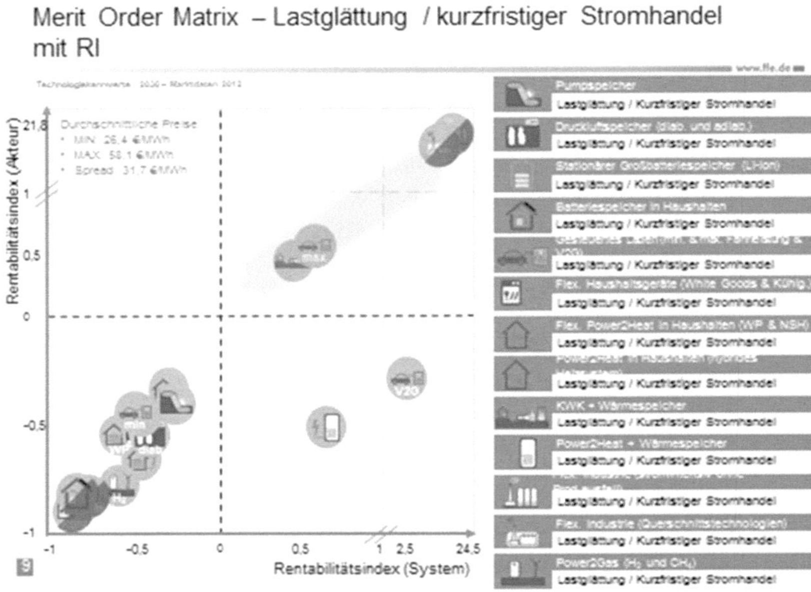

Bild 8 Merit Order Matrix für Funktionale Speicher

Kopplung des Energie- und Mobilitätssystems

Mit Blick auf die bisherige Entwicklung des Energiesystems und die bevorstehenden Herausforderungen an den Energie- und den Verkehrssektor lassen sich die wesentlichen Zusammenhänge beider Systeme in Bild 9 überschlägig verdeutlichen. Unterschieden wird hier zwischen den Teilsystemen für Elektrizität, Wärme, Kohlenwasserstoffen und der Energiebedarfsdeckung für die Mobilität. Über viele Jahrzehnte ging der Energiefluss im Bild ausschließlich von links nach rechts: fossile (und nukleare) Brennstoffe wurden für die Stromerzeugung in thermischen Kraftwerken eingesetzt. Später kamen Anlagen zur Kraft-Wärme-Kopplung hinzu, damit bestand eine Kopplung von drei Teilsystemen. Der Ausbau der Erneuerbaren Energien verbindet ebenso die Strom-, Wärme- sowie die synthetische Brennstofferzeugung und machte entsprechende Systemanpassungen erforderlich, z.B. den Ausbau von Leitungen und Speichern. Um negative residuale Lasten zu vermeiden und auch extreme temporäre und lokale Überschußerzeugung von erneuerbarer Elektrizität nutzen zu können, wird zunehmend der „Rückweg" von der Strom- zur Brennstoffseite (von rechts nach links) erforderlich. Hierzu zählt die Wasserstoffelektrolyse und die Methanisierung von Wasserstoff, um ihn in das vorhandene Erdgasnetz einspeisen und dessen Speichermöglichkeiten nutzen zu können.

Im Falle einer Wiederverstromung dieser Energieträger ist die Wirkungsgradkette sehr ungünstig. Effizienter ist daher die Nutzung dieser regenerativ erzeugten Energieträger im Verkehrssektor, mit einem doppelten Nutzen: elektrische Antriebe sind energetisch günstiger als Verbrennungskraftmaschinen und es lassen sich große Mengen an erneuerbarer Energie in den Verkehrssektor integrieren.

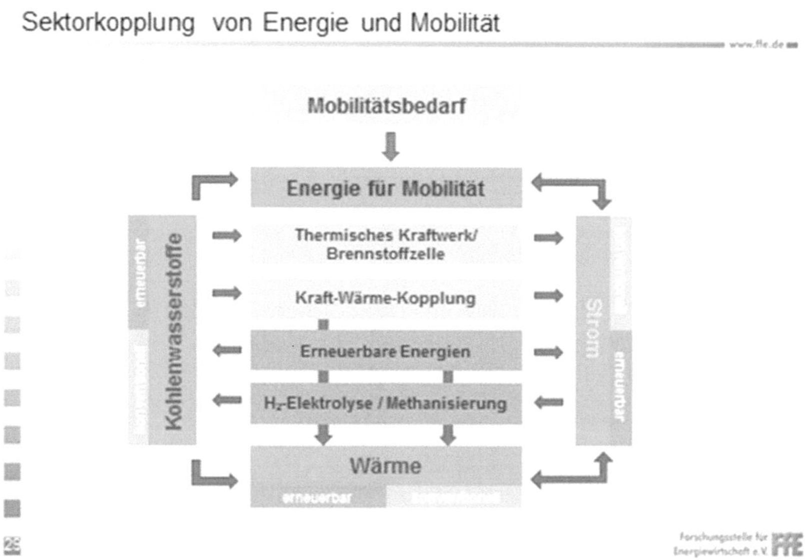

Bild 9 Die Kopplung zwischen Energie- und Mobilitätssystem

Fazit

Das Strom- und das Verkehrssystem haben sich bislang ohne wesentliche Verbindung parallel entwickelt. Es bestehen interessante Möglichkeiten und Notwendigkeiten zur zunehmenden Kopplung beider Systeme, um auch im Verkehrssektor die ehrgeizigen Klimaziele erreichen zu können.

Hohe Anteile an erneuerbaren Energien verursachen lokal und temporär große Überschussmengen. Innovative zusätzliche Stromanwendungen wie Elektrostraßenfahrzeuge (ESF) und Power2Heat (z.B. Wärmepumpen) können dieses Angebot flexibel nutzen und Abschaltungen regenerativer Erzeugungsanlagen zu Spitzenzeiten vermeiden helfen.

Ein ESF am Ladegerät kann zur Bereitstellung von positiver und negativer Regelleistung herangezogen werden, auch die Speicherkapazität eines Elektrofahrzeugs kann im Bedarfsfall für netzdienliche Aufgaben bereitstehen.

Eine technische und marktliche Kopplung von Energie- und Verkehrssystem bietet nach Einführung übergreifender Geschäftsmodelle noch erhebliche Potenziale zur weiteren Systemoptimierung (Flexibilisierung unter Einschluss innovativer Mobilitätskonzepte, Sektorkopplung mit industriellen Anwendungen, etc.).

München, 13.05.2016

Quellenauswahl

C. Pellinger, e.a.: MOS 2030 (Merit Order der Energiespeicherung), Forschungsstelle für Energiewirtschaft e.V, München, Mai 2016

U. Wagner, M. Rudolph: Energieanwendungstechnik, Springer-Verlag Berlin Heidelberg, 2008

U. Wagner: Vorlesung „Elektrische Straßenfahrzeuge" an der TU München, 2016

U. Wagner: Was bewegt uns in Zukunft?, Tagungsband „Neue Verkehrs- und Energiesysteme für umweltschonende Mobilität", ifo/Bayerische Akademie der Wissenschaften, Januar 2012

Synergies and challenges when transportation demand meets the electricity sector

Tetyana Raksha (Dipl.-Ing.)
Project Manager

Patrick R. Schmidt (Dipl.-Ing.)
Senior Project Manager, Partner

Colin Bendig-Daniels (BSc. Env)
Researcher

LBST – Ludwig-Bölkow-Systemtechnik GmbH
Daimlerstr. 15, 85521 Munich/Ottobrunn, Germany
T: +49 (0)89 608110-43
E: tetyana.raksha@lbst.de
W: http://www.lbst.de

Contents

Abstract

Scenarios on energy transition in transportation in Germany show that future renewable electricity needs will be substantial even when assuming high well-to-wheel efficiencies. On the supply side, integrating (fluctuating) renewable electricity into the electricity system poses its challenges. On the demand side, integrating certain electricity demand characteristics from transportation poses challenges of their own. This paper gives an overview over these challenges as well as the integration options specific to various electricity consumers in transportation.

Keywords: Renewables, Integration, Power-to-Gas, Transportation, Power Grid, Ancillary Grid Service.

1 Introduction

The transport sector is currently responsible for about 18 % of the greenhouse gas emissions in Germany (without maritime shipping and international aviation)and the energy industry for some 39 %. Today's predominantly fossil-based transport sector has so far developed very much independently of the remaining energy consuming sectors and could not reduce its greenhouse gas emissions in absolute terms. Unlike the energy and industrial sectors, the transport sector's greenhouse gas emissions have actually increased since by ~1 % between 1990 and 2014. The figures show the crucial role that an electrification of transportation on the basis of renewable electricity is playing if very high greenhouse gas reduction targets are to be successfully achieved.

For transportation, renewable electricity can either be used directly for charging batteries or to produce electricity-based fuels: power-to-gas (PtG) in the form of compressed hydrogen and methane, or power-to-liquids (PtL) as synthesised liquid hydrocarbon fuels, or both as part of 'power-to-anything' (PtX).

The necessary link between the electricity and the transport sector has received more attention in the past few years only. On the one hand, the new power demand from the transport sector is a challenge for the further development of the electricity system – in particular the closer the connection between the power supply and the electricity demand from the transport sector is as in the case of catenary power and fast charging stations. On the other hand, new possibilities are increasingly becoming available to optimize the entire energy system matchmaking between production and consumption – especially through greater use of fluctuating power generation systems (wind, solar) for the production of PtG or even PtL fuels.

The focus of this paper is to identify the growing demand for electricity from the transport sector and to assess its impact on the power grid.

2 Characterization of the Development of Total Electricity Demand

The substitution of fossil fuels by electricity and the use of electricity-based fuels for transport are expected to greatly increase demand for electricity.

The following figure shows possible options of power demand in Germany to 2050 for selected, and through the transport sector's increased power demand scenarios. All scenarios described therein are detailed modeling of the vehicle fleet, the driving performance and PtX production, which allows an informed assessment of the electricity demand. The scenarios focus is on road transport and inland navigation. Rail transport in Germany has generally been switched to electricity operation already. Air transport and maritime shipping is excluded. The transportation demand has been divided into a high and a low variant [LBST 2016]. The low transportation demand variant has been derived from the eMobil 2050 study's "Regional" scenario [Öko-Institut 2014]. The high transportation demand variant assumes further growth of cargo transportation as stipulated by the official traffic forecast 2030 of the Federal Transport Ministry and an extrapolation to 2050 as assumed in the context of the Mobility and Fuels Strategy (MFS) of the German Federal Government.

The demand of the transport sector in 2050 will range between 306 TWh/a in scenario "eMob+low" and 706 TWh/a in scenario "FVV+high". In case of the "FVV+high", electricity demands of the transport sector (excluding the sectors air transport and maritime navigation) would exceed the electricity demands of all other sectors combined (industry, private households, commerce and agriculture).

In scenario "CNG/LNG with $PtCH_4$", the production of synthesized methane for transport requires 614 TWh/a of electricity. In scenario "RE $PtCH_4$ + FCEV" the electricity demand is considerably lower due to efficient fuel cells and BEVs, namely some 447 TWh only.

Hydrogen production is assumed to be carried out with 100 % electrolysis in all scenarios. Electricity supply for PtX is assumed 100 % renewable in the scenarios. Today, hydrogen for transport use is already produced with about 50 % renewable electricity, albeit starting from low absolute amounts yet. For power-to-methane several demonstration and pilot plants have been put into operation in the last two years. With regards to PtL, first initial demonstration/pilot plants are in operation. Availability of larger-scale plants based on renewable electricity can be expected from 2020 to 2030.

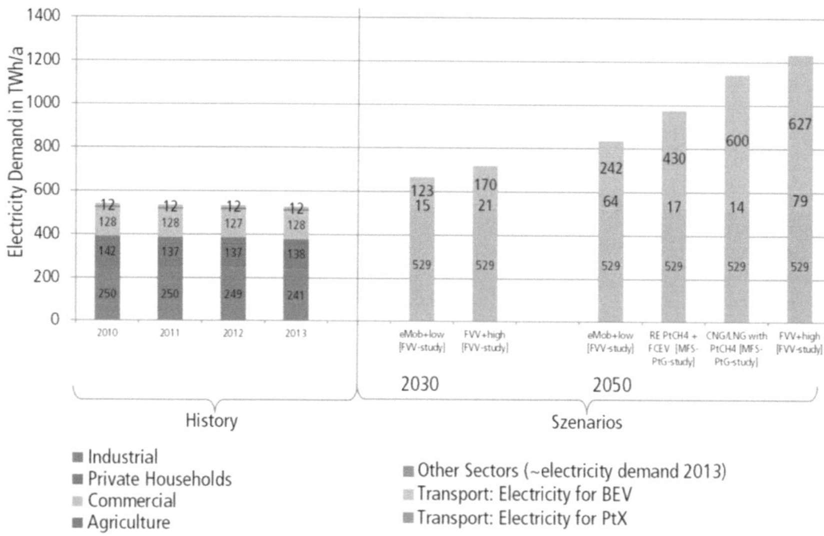

Development of the electricity demand 2010-2050 (image: LBST based on data [LBST 2016], [DLR et al. 2014], BDEW data)

Based on the presented scenarios, it can be concluded that the substitution of fossil fuels by electric energy and the additional use of power-based fuels will lead to a large increase in overall electricity demand. The transport sector will become a major electricity consumer (electricity for BEV and for electricity-based fuels (PtX)) by 2030-2050.

The additionally required renewable energy capacities for the provision of increased power demand of the transport sector are much greater than those currently presented in most scenarios for renewable energy expansion.

3 Network Restrictions – Characterization of the Effects of Electricity-Based Fuel Production on the Operation of Power Grids

Even today the German power supply system faces major challenges to integrate the growing demands for fluctuating renewable generation of the energy system. The need for an extensive network expansion both in transmission and in distribution networks has been identified.

Increasing electricity demands from transportation – as show in the preceding chapter 2 – further drives the need for additional expansion of the electricity grid, if integration is not well-done.

To explore a worst case scenario, a theoretical thought experiment of how the required infrastructure as a result of the power demands from the transport sector would present the worst case scenario. The thought experiment is based on the hypothetical assumption that all electricity demand from the transport sector in southern Germany (Baden-Württemberg, Bavaria) was supplied by offshore wind power from North Germany and transported to southern Germany via transmission lines.

Starting with the scenarios from the FVV study [LBST 2016], an additional power demand from the transport sector in southern Germany for 2050 is required from about 86 TWh in the "eMob+low" scenario to about 200 TWh in the "FVV+high" scenario. If this power were not locally produced, but is to be transported from northern to southern Germany it would require initial offshore wind[1] capacities of about 20 GW "eMob+low" to 44 GW "FVV+high". According to the theoretical thought experiment, these services would have to be installed by 2050 on electricity transmission lines. The transmission of large amounts of electricity over long distances is preferable via high voltage direct current transmission (HVDC). For this purpose either 5 or 11 HVDC overhead[2] or 10 or 22 HVDC cable lines[3] would be needed for the "eMob+low" or "FVV+high" scenario respectively, or combinations of overhead and cable lines. By comparison, the second draft of the network development plan 2025 provides for three north-south HVDC lines with a total transmission capacity of 10 GW [NEP 2016, scenarios B1/B2].

If the required hydrogen in scenario "FVV+high" is produced directly at the place where renewable electricity is generated and is transported by means of a hydrogen pipeline to the south, then a single hydrogen pipelines[4] would be sufficient. The direct electricity consumption in the transport would in this variant of the thought experiment of a two HVDC overhead line route.

In the best possible case (best case in terms of the lowest possible expansion needs for transport infrastructure), the regional electricity demands can be met by renewable electricity generating capacities in the region. The annual technical potentials for renewable power generation in Bavaria and Baden-Württemberg are between 150 and 220 TWh/a, derived from [DLR et al. 2014a]. The net power consumption in southern

1 Offshore wind capacity factor: 51 %.
2 Overhead power line: 4 GW (HVDC-VSC, ±500 kV=).
3 Cable lines: 2 GW (HVDC-VSC, ±525 kV=).
4 Pipeline: diameter 1420 mm, 10 MPa; H_2 transport capacity 40 GW.

Germany today is around 150 TWh/a. Hence, the regional technical renewable power production potential is not sufficient to cover the conventional and transport scenarios's renewable electricity demands in southern Germany.

There is principally a conflict between a higher demand for electricity grid expansion with its economic, political and environmental consequences and an optimized operation of the power systems.

Electricity-derived fuels (PtG methane, PtL) can be transported via the established infrastructures for methane and hydrocarbon liquids in Germany. Producing PtX nearby renewable electricity generation sites may reduce the need for additional power lines.

In addition to adequate transmission capacities the future power supply system must meet certain criteria regarding security of supply. A significantly higher ancillary services need from the German power grid is expected in the future. Renewable power generators and transport electricity consumers should take on tasks for grid stabilization (guarantee of the necessary system services), which today conventional power plants primarily provide.

Technical solutions and potentials for an alternative provision of ancillary services are available: The transport sector is potentially a flexible electricity consumer who can contribute significantly to bring supply and demand in accordance with the power system based on renewable energy and who can render the necessary ancillary services for the operation of the networks.

4 Characterization of Electricity Demands for Battery Electric Vehicles and for PtX Production

To determine the options and potentials for synergies between an increasingly electrified transport sector and an increasingly renewable electricity sector, it is important to first develop a profound understanding of the characteristics of individual power consumers who will become part of the transport sector.

BEV

Thanks to high well-to-wheel efficiencies, even an assumed strong increase in battery and catenary vehicles causes only a moderate rise in the annual electricity demand in Germany. However, power demand has to be thoroughly analyzed. The electricity demand profiles of the individual electrified vehicle groups differ considerably:

For example, charging battery electric vehicles in places where they are parked over a longer time (overnight in the garage or at work during the day) can be performed at relatively low power levels and a higher degree to match BEV slow charging with actual availability of renewable electricity.

Catenary-connected vehicles have the highest efficiency in converting grid power into propulsive energy, but also show the most direct temporal and spatial coupling of electricity use to the driving style. In order to make electricity consumption via catenary lines more flexible, additional measures are required, such as electricity storage on board or at the feed-in points of the catenary system.

PtX

The electrolysis plant for the production of hydrogen is the key component in the production of all electricity-derived fuels considered in this paper.

The introduction/use of power-based fuels can lead to a significant increase in the total electricity demand. In the case of FVV scenarios eMob+low and FVV+high, transportation electricity demands (PtX) are 242 TWh/a and 627 TWh/a, respectively. This is equivalent to some 42 % and 108 % of the net electricity production in 2014 (580 TWh).With PtX, both centralized and decentralized hydrogen production is possible. In the case of the central hydrogen production for renewable energy generation or at H_2 service stations pressure hydrogen (CGH_2) is generated and stored centrally via electrolysis. The hydrogen, obtained from a storage facility, can be transported, for example, through a pipeline network to service stations. The electrical connected load of the electrolyser in this case is about 20 MW or more; the systems are connected to the high-voltage network ($\geq$ 110 kV). Decentralized hydrogen production is characterized by smaller electrical loads and thus connected to the medium voltage to lower voltage levels (e.g. 10 kV and 0.4 kV respectively).

The PtX fuel production is not only technically possible at almost all plant sizes but is also scalable. A special role of storage in implementing PtG has to be emphasized. Through a hydrogen storage, hydrogen production (electricity demand) and hydrogen demand is temporally or spatially decoupled. An intermediate storage of hydrogen is required in almost applications anyway.

5 Technical Potential of New Electricity Demand from the Transport Sector for the Provision of Ancillary Grid Services

Possible new power demands from the transport sector are a challenge for the electricity system as it stands today. On the other hand, new options are increasingly becoming available to optimize the entire energy system through further integration of production and consumption. PtX fuel production and e-mobility can contribute to network relief and consumption level increase plus enabling middle to long-term, central and decentralization network system services.

The following figure gives an overview of technical options of new electricity demand from transport to reduce or even solve challenges associated with increasing power demands from transport and increasing renewable power generation in the electricity grid.

Ancillary Services → / ↓ Options		Frequency Control	Reactive Power / Voltage Control	System Restoration	System Control
Electricity directly (BEV / Overhead line):					
BEV (slow charging)		+	++	+	
BEV (fast charging)			+		
Electrified trucks (Overhead line)		+			
Electrified busses			+		
Production electricity-based fuels (PtX):					
PtG-H_2 for FCEV					
	Centralized	+++	+++		
	Decentralized	+++	+++		
PtG-CH_4 for ICE					
	Centralized	+++	+++		
	Decentralized	+++	+++		
PtG-CH_4 / PtL for ICE					
	Centralized	+++	+++		
	Decentralized	+++	+++		

↓ Storage density (System Restoration column); ↓ Power per unit increases (System Control column)

LBST, 2016-04-22

+++ very suitable; ++ suitable; + possible in principle

Technical potentials of new electricity demand from the transport sector for the provision of ancillary services (image: LBST)

Whether and to what extent the possibilities referred to in the figure can be developed and used for the provision of ancillary services depends heavily on the future technical, economic, regulatory developments in the energy system.

To summarize the characterization of the power demand from transport it can be stated that the various fuels/drives have opposite advantages and disadvantages in terms of their efficiency in energy supply (upstream processes) and use (drive) as well as in their flexibility for electricity generation, and not least in terms of energy storage density of fuels. The subsequent figure shows an overview of the qualitative relationships.

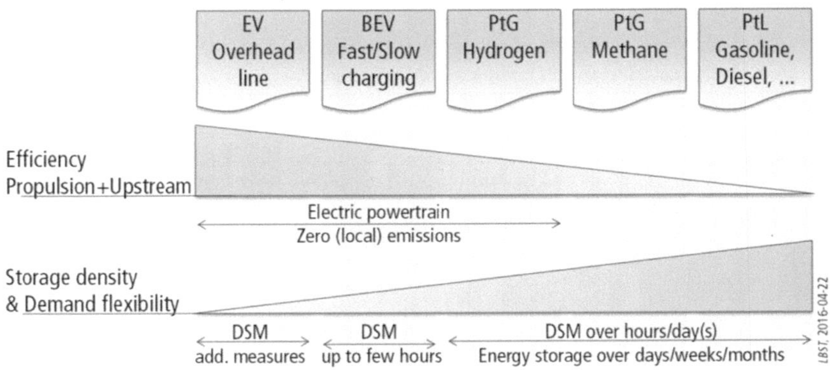

Efficiency, storage density and efficiency of different fuels/drive (image: LBST)

Above figure shows that there is a trade-off relation between the different fuels/drives, in particular between the efficiency and the degree to support renewable power integration into the system. Without additional measures, such as electricity storage or hybridization, overhead and BEV charging systems tend to be harder coupled to the power grid. PtG methane and especially PtL fuels for use in combustion engines are less efficient 'well-to-wheel', however, they can provide high flexibility in responding to the availability of renewable power in the grid. It has to be noted that hydrogen represents a robust compromise for the conflicting priorities. In fact, hydrogen is the only fuel/drivetrain option providing both local zero emission capability (due to the electric drivetrain) and long-term electricity storage capability (because it's a chemical).

Electricity-based gaseous fuels are a versatile technology that can be used across sectors and systems. PtG facilities (electrolysers) are potentially flexible loads, with each unit being significantly bigger and more flexible than BEV, thus reducing system complexity. PtG use in mobility requires the construction of a new hydrogen filling station infrastructure or the adaptation of the existing methane refueling infrastructure. The PtG distribution infrastructure is of great importance since with its support the fuels are not only transported from the production site to the consumers, but the distribution infrastructure can also take over the storage function of the temporal and spatial decoupling of the fuel production and use. Both centralized and decentralized infrastructure concepts offer a synergistic use is possible through a combination of PtG as an electricity storage facility and as fuel. Both centralized and decentralized infrastructure concepts offer a synergistic use of PtG as electrical storage and fuel.

The power-to-gas technology is currently not commercially viable. In general, it is conceivable to use PtG not only in the transport but also in the electricity and gas market or manufacture. The revenues to be generated through these synergies can significantly improve the efficiency depending on type of use, utilization and sizing and open up new business. A cost reduction for the benefit of all sectors involved is due to the larger number of installed systems (quantity effects) as well as to increasing plant sizes (capacity effects). The regulatory framework has a significant impact on the economic operation of PtG systems and should be adjusted accordingly.

6 Conclusions

From a long-term perspective an energy transition in the transport sector is in principle feasible. With 1000 TWh per year the technical potential for the generation of renewable electricity is significant in Germany. It can roughly cover prospective transport's renewable electricity demands in Germany, albeit not yet including maritime shipping and international aviation.

Transport potentially represents a flexible demand for electricity, which can contribute significantly to match supply and demand in a renewable energy power system and to provide the necessary ancillary services for the operation of networks. Important potential contributions of transport to the electricity system in energy transition are:

- Integration of short-term renewable energy peak supply by flexible charging of battery vehicles;
- Flexibility of electricity demand through flexible fuel production of hydrogen, methane or synthetic liquid fuels;
- Provision of balancing power to stabilize the power systems through battery-powered vehicles, PtG-, and PtL-plants;
- Provision of reactive power (in principle in all systems possible which are connected via an inverter to the power grid);
- Provision of black-start capability to restore the power system after a grid failure,
- Reducing the need for the power grid expansion through intelligent coupling of electricity and gas networks as well as infrastructures for liquid fuels; and
- Providing dispatchable generation capacity (secure power) through re-electrification of PtG/PtL where necessary.

To exploit these potentials for the energy transition, the future transport demand for electricity should be considered in the network development plan or in the revisions to the Renewable Energy Law (EEG) and other relevant regulatory frameworks. In particular, additional electricity demands from transport have to be matched with additional deployment of renewable power plants, for example, through a corresponding increase of renewable energy expansion plans/targets.

Regulatory frameworks have a significant impact on the economic operation of centralized and decentralized PtG plants. Currently, there are regulatory uncertainties and disadvantages, particularly as regards taxes or exemption.

The problems associated with rising electricity demand from transportation on the electricity system are not predictable over the time-span of several decades; robustness is rather given with short- to medium-term developments. Readjustments have to be weighed, on the one hand, between planning and investment security for all concerned, and on the other with regard to flexibility and adaptability to changing social and political conditions.

Acronyms and abbreviations

BDEW German Association of Energy and Water Industries

BEV battery-electric vehicle

CGH_2 compressed hydrogen

CH_4 methane

CO_2 carbon dioxide

DSM demand-side-management

EEG Renewable Energy Law

eMob electric mobility

FCEV fuel-cell electric vehicle

HVDC high-voltage direct current

ICE internal combustion engine vehicle

MFS Fuel and Mobility Strategy of the Federal Government

PtG power to gas

PtL power to liquid

PtX power to anything, electricity-based fuels

REEV range-extender battery-electric vehicle

RES renewable energy source

WtW well-to-wheels

Literature

DLR et al. 2014: DLR, IFEU, LBST, DBFZ: Power-to-Gas (PtG) im Verkehr – Aktueller Stand und Entwicklungsperspektiven; study commissioned by the Federal German Transport Ministry (BMVI) for the German Mobility and Fuels Strategy, MAY 2014

DLR et al. 2014a: DLR, IFEU, LBST, DBFZ: Erneuerbare Energien im Verkehr – Potenziale und Entwicklungsperspektiven verschiedener erneuerbarer Energieträger und Energieverbrauch der Verkehrsträger; study commissioned by the Federal German Transport Ministry (BMVI) for the German Mobility and Fuels Strategy, OCT 2014

DLR et al. 2016: DLR, IFEU, LBST, DBFZ: Analyse von Herausforderungen und Synergiepotenzialen beim Zusammenspiel von Verkehrs- und Stromsektor; study commissioned by the Federal German Transport Ministry (BMVI) for the German Mobility and Fuels Strategy, ongoing investigation

LBST 2016: Schmidt, P., Zittel, W., Weindorf, W., Raksha, T. (LBST – Ludwig-Bölkow-Systemtechnik GmbH): Renewables in Transport 2050 – Empowering a sustainable mobility future with zero emission fuels from renewable electricity – Europe and Germany; ed.Forschungsvereinigung Verbrennungskraftmaschinen e.V. – FVV, JAN 2016

LBST 2016a: Schmidt, P., Weindorf, W. Raksha, T., Zerhusen, J., Michalski, J. (LBST – Ludwig-Bölkow-Systemtechnik GmbH): Power-to-gas – Short term and long term opportunities to leverage synergies between the electricity and transport sectors through power-to-hydrogen – Application A: Hydrogen from power-to-gas for use in refineries; An expertise for the Fondation Tuck, 2016

NEP 2015: Szenariorahmen für die Netzentwicklungspläne Strom 2015. Entwurf der Übertragungsnetzbetreiber. 30 APR 2014

NEP 2016: Übersichtskarten des zweiten Entwurfs NEP 2025; FEB 2016

Öko-Institut 2014: Öko-Institut e.V.: eMobil 2050 – Szenarien zum möglichen Beitrag des elektrischen Verkehrs zum langfristigen Klimaschutz; ed. BMUB, Berlin, 2014

OVERALL ENERGY SYSTEMS I

Grid integration of electric vehicles

Xaver Pfab, Victoria Haese

BMW AG, Munich

1 Why grid integration of E-mobility?

Electric mobility has been an integral part of the BMW product strategy since the introduction of the BMW i models i3 and i8 and the expansion of the model variety by various plug-in-hybrid vehicles. Beside of changing insights of customers worldwide in terms of future mobility and growing importance of sustainability, all OEM are faced to huge challenges in fulfilling the CO2 emissions legislation in the main markets worldwide. For premium-OEMs as BMW this means an extra effort due to the wide model program also in the sportive and luxury segment. Therefore, BMW has identified the electric vehicle as a necessary approach for the future success of the company.

To make E-Mobility in total a success story, it needs several solutions:

1. Attractive vehicles according to customer's requirements, highlighted: range, short charging time, hassle-free handling for charging, affordable prices and cost effectiveness under a total cost of ownership's view
2. A charging infrastructure that leave no doubts in terms of the next available charging station including enough fast-charging stations along the routes.
3. Full compatibility to grid standards, energy market legislation and constraints. E-Mobility powered by a maximum of renewable sources.

The third point stands for "grid integration of E-Mobility". This new item – according to BMW's understanding covers:

- Emission free, sustainable Mobility: Provide a complete power supply from renewable energy to E-Mobility to achieve a robust well to wheel approach
- Grid-infrastructure is consistent with E-Mobility: Identification of grid-constraints by growing E-Mobility and early development of measures to avoid or reduce inappropriate investment into grid refurbishment
- BMW and BMW-customers have benefits due to new business-models: Identification of new business opportunities by using the flexibility based on the real charging time versus the parking time being plugged-in and the vehicles storage capacity.

By analyzing customer groups and major use cases, two main situations can be identified: First, the electric vehicle is parked at home and effects from charging to the household installation behind the meter have to be considered: in the minimum scenario with any other unconventional consumers as heat pumps, that may already make sense for an intelligent load or energy management and any photovoltaic installation, the only check point is the capacity of the wiring and the fuse of the charging point. In the maximum scenario, including all that and even a battery storage in combination with a PV system, topped by a home energy management system (HEMS), the dominant question is: "How to integrate electric vehicle charging into a local grid to achieve a maximum of self-consumption by a minimum of costs?"

Second, having a look into the grid beyond the meter, effects from only one single electric vehicle are marginal. Migrating to higher populations of electric vehicles, there is a change from some single kilowatts of charging power to cumulated megawatts. This arises questions in terms of local grid capacity and perhaps needs for grid refurbishment or alternative measures in a way of a grid load management with the perspective of the need to achieve control on the charging processes. As there is potential for a significant flexibility while charging, the next question is about opportunities to get into the use of that flexibility in a beneficial way, not only to utilities and grid operators, but the electric vehicle customer as well.

2 Key figures E-Mobility vs. grid and power generation

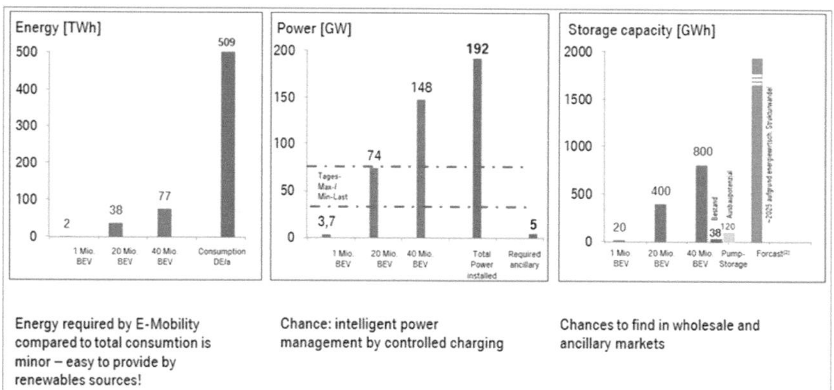

Figure 1: E-Mobility and electric power supply, key figures Germany. Maximum figures without usage factors.

Sources: Bundesministerium für Wirtschaft, Zahlen und Fakten Energiedaten (Brutto-Stromverbrauch), 02/2013
BDEW: installed power plant capacity 2014
Auer, Deutsche Bank (DB) Research: Moderne Stromspeicher, Unverzichtbare Bausteine der Energiewende, Jan 2012 . Speicherbedarfsprognosen.
Premises electric vehicle: standard charging power 3,7kW; energy consumption per year: 1920kWh (@12.000km per year);
Usable storage capacity per vehicle assumed for daily ancillary services: 20kWh with future battery technologies after 2020.

Compared to the general power generation and consumption in Germany, the required amount of energy for E-Mobility appears quite small. This leads to two conclusions: There is no significant need in terms of additional power generation due to a secure supply. To fulfil the requirement of a complete, sustainable and emission-free power supply for E-Mobility, no excessive investment for additional wind- and photovoltaic-power plants would be necessary. Even from an customers point of view, energy cost of electricity and even of renewable electricity are below of the actual very low fuel costs of gasoline and diesel.

Power peaks from cumulated charging events as shown in the second diagram of Figure 1, are also not that critical in terms of the overall grid stability at least on the transportation grid level and in the nearer term. However, in a longer term the situation has to be considered in a more detailed approach: coming to bigger populations of E-Mobility in certain locations, the distribution grid level limitations may become visible in some cases. As planning schedules for electricity grids usually have a timing horizon of 15 years and more, it's the right time now to consider the impact of E-Mobility on the electricity grids. Beside of the negative aspects of grid overload in time slots with charging power peaks, we also see a great positive aspect in the opportunities of an intelligent controlled charging.

Charging events have a high amount of flexibility, as vehicles are plugged-in usually over night with more than 10 hours, while typically only three hours are needed to fully recharge the battery. This gives space for various intelligent approaches. By shifting the charging event into advantageous time-slots, a "grid-friendly" behavior could be achieved. Simultaneously the customer has the possibility to save money by using cost-effective time-of-use tariffs, or even earning revenues from grid services when pools of many electric vehicles are under the charging control system of a virtual power plant acting in markets as demand response or ancillary services.

With growing shares of the renewable sources as wind- and photovoltaic-power, power generation becomes further more volatile, with times of under- and over-capacity. In accordance to that, flexibility and storage capacity will become important values in the future electricity system. As the third diagram in Figure 1 shows, E-Mobility can help with both, even with todays charging flexibility. In the mid-term future it will be possible to develop sustainable business models, with a significant benefit to every stakeholder involved, for complete system solutions like V2H ("vehicle to home") and V2G ("vehicle to grid"). This could actually be reached by using the electric vehicles batteries in a reverse-charging manner.

3 Electric vehicle customer premises and requirements

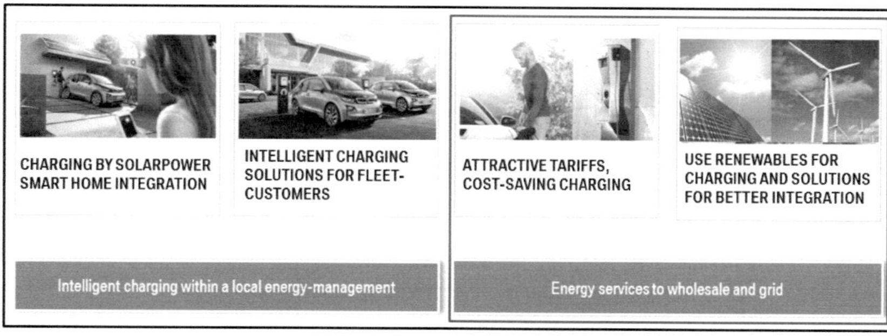

Figure 2: Customers requirements depending on their domestic situation

Intelligent charging processes that exceed the capabilities of standard (immediate) charging can only be launched on the market if they satisfy crucial customer needs.

In a further contribution of this conference, "Local Energy Systems for Electric Mobility" (Dr. Willibald Prestl, Bernhard Brendle, Dr. Michael Beer, BMW AG, Munich), a detailed analysis of the typical todays situation of BMW i3 customers is shown: The majority is living in an own house, parking and charging at home is mandatory, 71 % use green electricity, 45 % use photovoltaics for domestic energy generation – perfect prerequisites for intelligent charging within a local energy management.

As E-Mobility has to become a convincing offer to far wider customer groups, additional energy services to wholesale markets and the electricity grid (Figure 3) can be a new way for vehicle customers to save energy costs, and even being an active part of the future distributed, sustainable energy system as well.

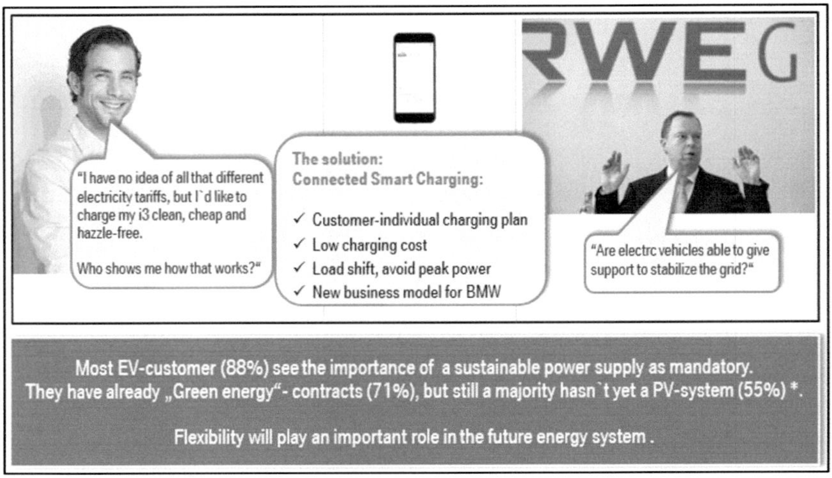

Figure 3: Use cases smart charging: Beside the vehicle customer, new stakeholders will appear.

*) Source: PREMIUM –Project (Plug-In-, Range-Extender- und Elektrofahrzeuge unter realen Mobilitäts- Umständen: Infrastruktur, Umweltbedingungen und Marktakzeptanz), BMW AG et alt.

4 Energy transition & E-Mobility: a strong interference

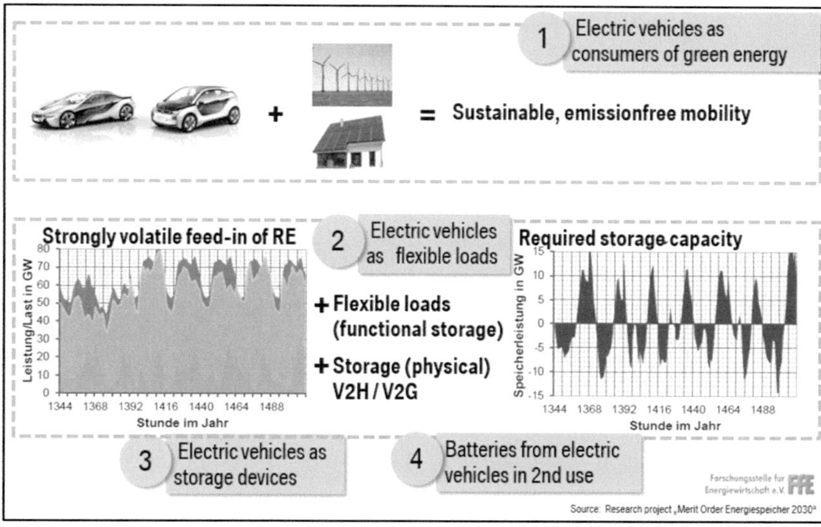

Figure 4: Steps of interaction between the electricity system and E-mobility

The first step of interaction between E-Mobility and the electricity grid is – of course – drawing energy while charging. To fulfil the demand for an emission-free and sustainable mobility, charging with power from renewable sources is mandatory. Critics of E-Mobility usually complain that electric vehicle CO2 emissions are – calculated on the average CO2-figure per kWH electricity – very close to the CO2 emission levels of modern high efficient ICE vehicles.

As shown above, more than 70% of BMW i3 customers are aware of that and have already done their decision by owning a "green energy" tariff. And this is a lot compared to only 22% of German households having done that too. Not to forget the 45 % i3 customers using photovoltaics for domestic energy generation today. So it's in the customers opportunities and responsibility to achieve a real emission-free and sustainable mobility already today.

Beyond of these simple approaches in terms of green energy, advanced future functionalities are possible: if a certain information forecast signal for a high overshooting wind- and photovoltaic power generation is available, intelligent charging is able to shift charging events just into the right time slots. This was already demonstrated in one of our former research project "Gesteuertes Laden V2.0" (2010-2012) funded by the German BMUB.

The second step will be the use of the flexibility from charging events. Beyond the local use cases as local load management within fleet charging situations, a load drop requested by the DSO according to §14a, ENWG (Energiewirtschaftsgesetz), can reduce the customers grid fee significantly under today's German regulatory framework. Typically this is used with heat pump systems where an interrupt of even some hours is – due to the high storage capacity of a heating system – not even noticeable to the customer. So, this approach certainly need some adoptions when being used for electric vehicle as well, but huge hurdles are not evident.

The amount of flexibility of an electric vehicle is mainly defined by the charging power, the daily required energy capacity and the time being plugged-in. From various projects, where the user behavior was analyzed under multiple customer conditions, condensed figures are:

- Average daily driving: ca. 45km
- Average daily energy consumption: ca. 10kWh
- Average daily charging time: 3h (@ 3,7kW charging power)
- Average daily plug-in duration: 10h
- Shifting potential: ca. 7h

Several studies analyzing customer's charging behavior found: dominant is "charging when necessary" which means, that in average only 4.2 charging events per week will occur. BMW identified that the flexibility can be increased even more, when customers get information and incentives to optimize their charging behavior in a way to plug-in their electric vehicle more frequently, e.g. every night. This was tested and proven in the research project "Controlled Charging V3.0" (see also the contribution "Challenges and results in the BMUB research project "Controlled Charging V3.0" (Michael Westerburg et alt., EWE AG)).

With the capability of reverse charging and the opportunity to use an electric vehicle not only as a flexible load, but also as a real storage device, in a third step the amount of flexibility and the value out of it can be improved significantly. Very first electric vehicles with reverse charging capability are already offered. However, without a complete vehicle to home (V2H) or vehicle to grid system (V2G), consisting of the vehicle, a reverse flow wall box, a control system that manages the charging / discharging process, and finally an interface to electricity wholesale or grid services markets, no significant customer value can be generated.

Step four aims beyond the lifetime of an electric vehicle: Batteries from electric vehicles might be useful for stationary purposes for a longer time, if the vehicle has been damaged by an accident with a remaining proper battery, or the battery capacity has passed a certain capacity or SoH threshold, where guaranteed performance or range figures cannot be provided anymore. Most OEM's have projects to analyze potentials and resulting business cases based on such an approach.

5 Flexibility in charging and how to use it

As already shown, electric vehicles provide, next to the increased flexibility while charging, also a concept for individual mobility for people using their car 1 – 2 hours on average per day. The remaining 22 hours passenger cars are parked. However, one of the great advantages of the car in terms of individual mobility is the permanent availability to the owner or user. No planning efforts necessary at any time, that's a great benefit to the customer, which has to be considered and respected when creating new ideas how to use electric vehicles in the electricity system.

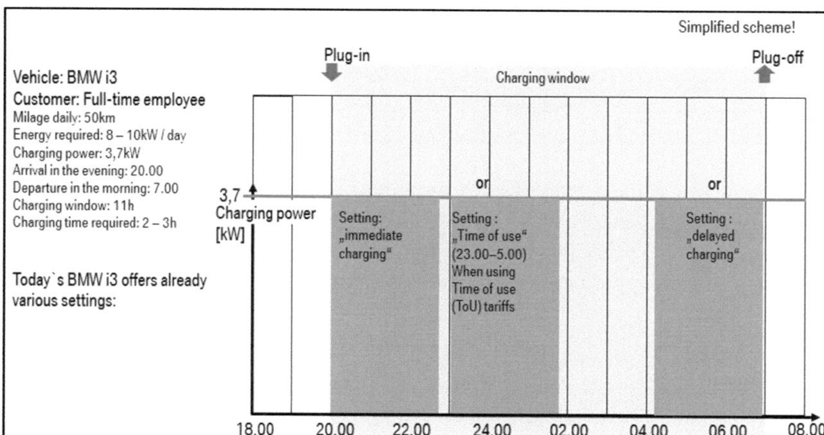

Figure 5: Use-case: „charging @ night": charging settings today

Already today, the BMW i3 offers to his customers some first opportunities in charging flexibility. Beside the standard setting of "immediate charging", the customer can also select a charging timer to adopt to time slot with lower electricity rates as "Time of use" tariffs (ToU) offered by utilities. This is the first possibility of saving money while charging. The saving potential varies strongly between different countries and utilities. Some snap shots are shown in table 1 for the German market with an average saving potential of 93€ or 13% of the charging costs. In other markets ToU-tariffs seem to be more attractive: in the USA, average, non-representative savings are 110$ or 32%, in UK 225£ or 50%.

Table 1: HT/NT tariffs in Germany and savings for electric vehicle charging

	Provider	Name	Schedule	Basic charge	energy charge		Energy-charge for 2800 kWh (100% OffPeak) + increased basic charge	Savings	Savings [%]
München	SWM	Mkompakt	immer	68,43 €	0,2532 €		708,96 €		
Q: SWM.de: 24.04.13		Mkomfort	Off-Peak	93,06 €	0,2211 €	-0,0321 €	643,71 €	**65,25 €**	**9**
Heidelberg	SW Heidelberg	Strom fix 14, M	Off-Peak	123,12 €	0,2270 €		635,60 €		
Q: SW HD.de 24.04.13		Strom fix 14 Nacht	immer	160,08 €	0,1808 €	-0,0462 €	543,20 €	**92,40 €**	**15**
Stuttgart	EnBW Vertrieb	EnBW Komfort HH	immer	94,53 €	0,2842 €		795,76 €		
Q: Vervivox		EnBW Komfort HH NT	Off-Peak	121,24 €	0,2187 €	-0,0655 €	639,07 €	**156,69 €**	**20**
Berlin	Vattenfall Sales	Berlin Kompakt Privatstrom	immer	70,80 €	0,2688 €		752,64 €		
Q: Verivox		Berlin Easy Privatstrom Spar Aktiv	Off-Peak	106,80 €	0,2417 €	-0,0271 €	712,76 €	**39,88 €**	**5**
Düsseldorf	SW Düsseldorf	Düsselstrom Klassik	immer	69,02 €	0,2594 €		726,32 €		
Q:Verivox		Düsselstrom Klassik HH Schwachlast	Off-Peak	138,04 €	0,2063 €	-0,0531 €	646,66 €	**79,66 €**	**11**
Köln	RheinEnergie	FairRegio Strom Basis	immer	84,25 €	0,2725 €		763,00 €		
Rhein-energie.com		FairRegio Plus Nebenzeit Option	Off-Peak	144,00 €	0,2472 €	-0,0253 €	751,91 €	**11,09 €**	**1**
Hamburg	Vattenfall Sales	Hamburg Kompakt Privatstrom	immer	73,20 €	0,2652 €		742,56 €		
Q:Verivox		Hamburg Easy Privatstrom Spar aktiv	Off-Peak	106,80 €	0,2381 €	-0,0271 €	700,28 €	**42,28 €**	**6**
Leipzig	Stadtwerke Leipzig	Strom21.online Privat	immer	124,95 €	0,2642 €		739,76 €		
Q: Verivox	AleMannen Strom	AleMannen Strom Tag und Nacht SIC	Off-Peak	129,60 €	0,2043 €	-0,0599 €	576,69 €	**163,07 €**	**22**
Erfurt	SW Erfurt	SWE Strom.	immer	129,96 €	0,2434 €		681,52 €		
Q: Verivox		SWE Strom. Total	Off-Peak	142,80 €	0,1966 €	-0,0468 €	563,32 €	**118,20 €**	**17**
Dietersheim bei Nü	N-Ergie	Strom Standard HH	immer	85,32 €	0,2775 €		777,00 €		
Q: Verivox		Strom Extra HH	Off-Peak	134,94 €	0,2226 €	-0,0549 €	672,90 €	**104,10 €**	**13**
Hahnenklee	Harzenergie	Basisstrom HH	immer	63,19 €	0,2749 €		769,72 €		
Q: Verivox		Schwachlast Strom HH	Off-Peak	78,54 €	0,2139 €	-0,0610 €	614,27 €	**155,45 €**	**20**
Hamburg	Lichtblick	Basisstrom HH	immer	107,40 €	0,2644 €		740,32 €		
Q: Lichtblick		§14a basiert - Autostrom_Pilot	21.00-06.00	23,40 €	0,1944 €	-0,0700 €	460,32 €	**280,00 €**	**38**

A new model is offered by Lichtblick using the German regulatory frame work of §14a EnWG (Energiewirtschaftsgesetz), where the customer allows his DSO to reduce load in critical grid situations. This can lead to further savings to the customer if he accepts a limitation of the charging into a certain time slot from 21.00 night to 6.00 in the morning.

Another option is to set a departure time, e.g. next morning, 7.00 o'clock. In this case, the BMW i3 is going to calculate a charging schedule where charging will start at a latest time to complete charging until departure, usually in combination with the preconditioning of the passenger cabin.

In the near future BMW will offer further options. In some countries "dynamic electricity tariffs" are under discussion or in development. If dynamic prices appear on the market, functional solutions can be offered by following the charging schedule to a price forecast signal as shown in Figure 6.

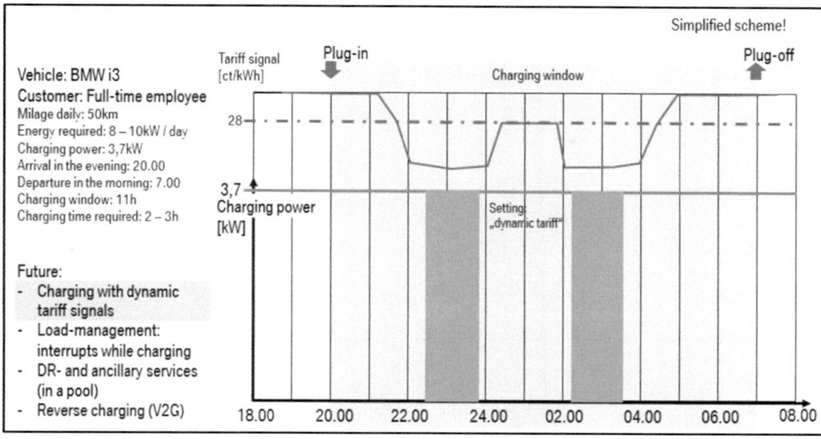

Figure 6: Charging in combination with dynamic price signal

Therefore, a permanent communication and interaction between the electric vehicle and a backend system for receiving price signal forecasts becomes mandatory. The system receives data from energy economy based partners and generates charging schedules for individual electric vehicles.

Another use case will occur in charging fleets of electric vehicles at one location. The vehicles could either be parked at a facility "behind the meter" as company parking lots, parking facilities in compounds, or even in a DSO area, where the maximum peak load from all given vehicles rises the grid or installation capacity. It may be more economic to install an intelligent load management system than to make an investment for a grid or installation reinforcement. This is motivated in avoiding high investment to align the maximum capacity to a single peak load during the day – occurring with a very low probability.

Concepts and first solutions are presented in the contribution of this conference, "Local Energy Systems for Electric Mobility" (Dr. Willibald Prestl, Bernhard Brendle, Dr. Michael Beer, BMW AG, Munich).

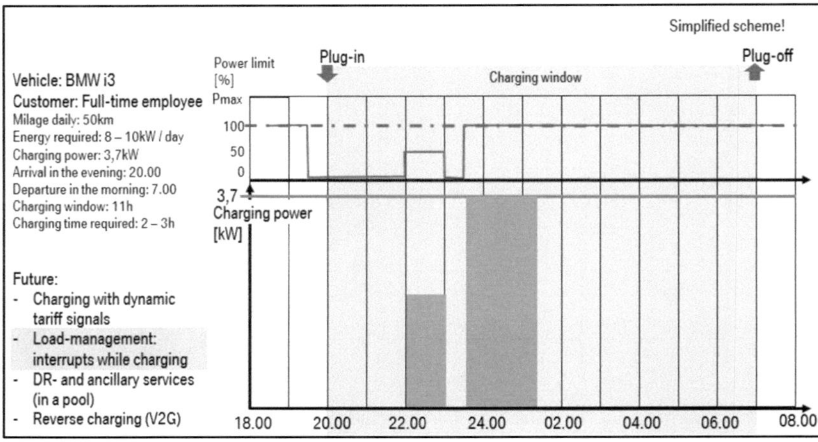

Figure 7: Charging in combination with a load management

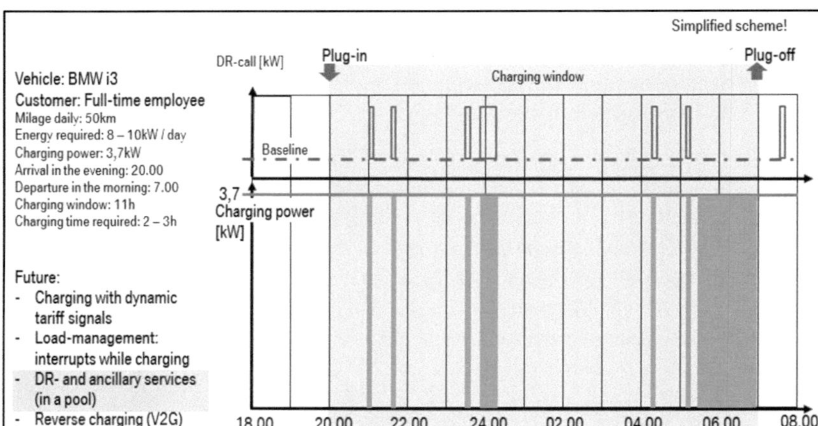

Figure 8: Charging of an electric vehicle within a pool of electric vehicles forming a "virtual power plant"

The most advanced perspective of using electric vehicles charging flexibility is the vision of a "virtual power plant" with really huge populations of electric vehicles. The single vehicle would face a charging schedule as shown in Figure 8.

The first hurdle that has to be overcome is convincing customers to participate in such business models. Today and in the next years as well E-Mobility is brand new and unknown. Main concern is about sufficient range and charging capabilities. The idea of a third party, managing charging events for a customer, might be seen as an additional challenge and a source of further uncertainty. As a result everybody, who is willing to come up with such solutions, has to provide a business model which convinces customers by showing up significant benefits, robustness and guaranteed charging at any time. Beyond of that, data security will be an issue of highest interest – not only in Germany and the EU.

Technical solutions for an electric vehicle pool control system are ambitious with respect to quite low delay times and absolute reliability as required for ancillary services. Nevertheless, BMW has learned from various projects how to fulfil these requirements with affordable solutions. One key element is a safe and protected communication path into and out of the vehicle. As many sub-functions of a pool control cannot implemented within vehicle's E/E-architecture, powerful backend systems as in the case of BMW's "Connected Drive" are mandatory system components. This is BMW's approach to enable a maximum of valuable and reliable grid integration functionality without any limitations in terms of an OEM's product liability.

6 Pathways into the markets and commercial use

Compared to the automotive industry, the energy economy is completely different. Figure 12 shows main structures and quantities of players only in Germany and the USA.

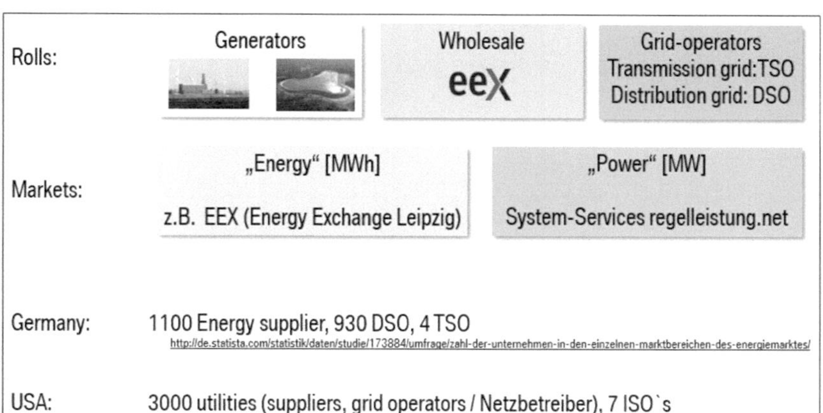

Figure 9: Energy economy: key indications

In the EU and in Germany there is a “liberalized market” with a strict separation of the main roles as generators, grid operators and wholesale. In the USA, in various states conditions are the same or similar, in others utilities are vertically integrated covering all mentioned roles, having local monopoles. The result are very different market access and products. For an automotive OEM very challenging.

To get a first idea of viable opportunities for the use of electric vehicles charging flexibility, BMW started three “Deep dives” as shown in Figure 10. Within the German “Showroom E-Mobility” program of the German government, BMW together with 6 further partners demonstrated by “Controlled Charging V3.0” how to generate ancillary services as “Secondary response” (Sekundär-Regelleistung) from a pool of electric vehicles. Results from this project are presented in this conference in the contribution “Challenges and results in the BMUB research project “Controlled Charging V3.0” (Michael Westerburg et alt., EWE AG) and “Smart charging in daily routine – expectations, experiences, and preferences of potential users” (Franziska Schmalfuß, et alt., Institute of Psychology, TU Chemnitz).

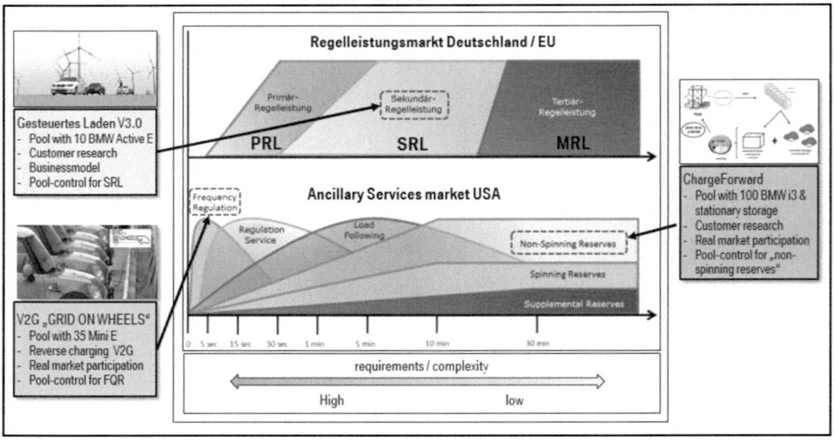

Figure 10: Evaluation of future markets, products and combinations with research projects

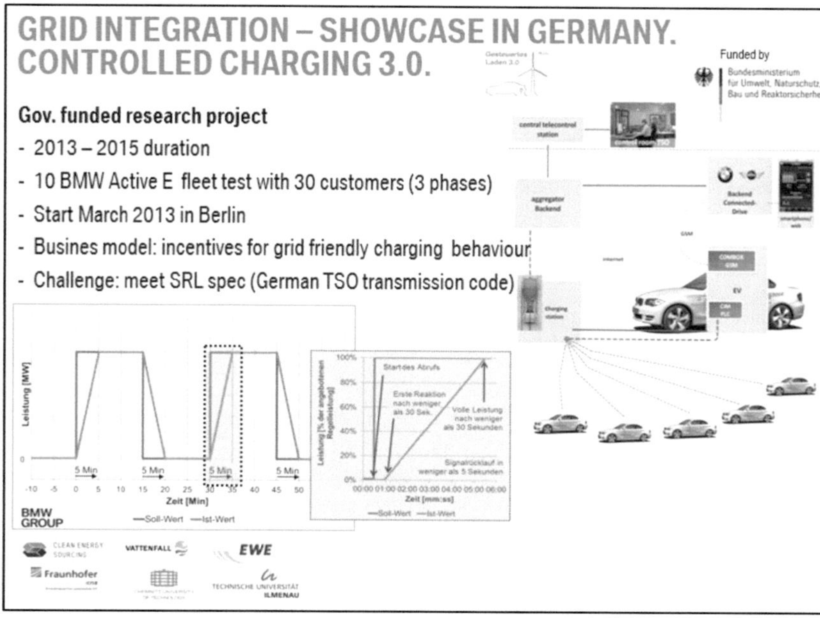

Figure 11: “Controlled Charging V3.0: demonstrator for secondary response from electric vehicles.

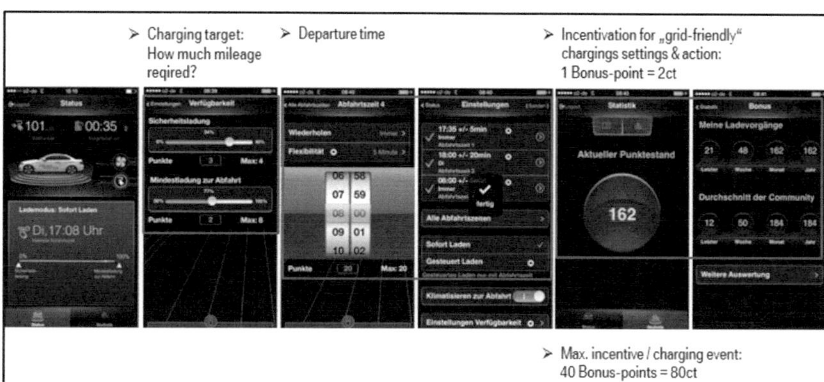

Figure 12: “Controlled Charging V3.0: User interface and incentives

During a 15month pilot operation, the system showed the capability to fulfil the Transmission Code specification of the German Transmission system operators. However, only using the charging events, no positive business case is in sight, due to the very limited charging time and capacity and to the falling revenues in the ancillary services market at least in Germany. V2G as shown in Figure 13 may bring better perspectives.

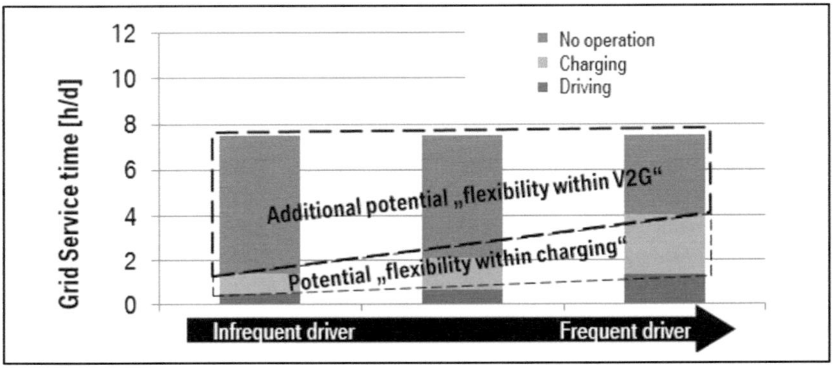

Figure 13: V2G can improve the potential in terms of ancillary services significantly

With participating in the "Grid on wheels" project, a consortium of the University of Delaware, PJM, NRG, Autoport, EV Grid and Princeton Power, a V2G demonstrator with a pool of 35 MINI E was built up and is now in a real frequency control operation since two years. From this project very valuable experiences could be gained:

- It is feasible to fulfil the technical requirements even from the fastest ancillary service as frequency regulation and to follow power signals for a whole pool of electric vehicles with the needed accuracy
- In the frequency regulation with fully V2G operation, cycling of the battery will exceed by factor 2 – 3 versus regular driving & charging. However, the depth of discharge (DoD) by V2G are quite small compared to driving conditions. The main concern, that V2G and reverse charging is a highly negative factor to battery lifetime and quality, could not be confirmed in this project. After two years of operation and a vehicle life of the MINI E of meanwhile 7 years, no significant aging was visible up to now. However, MINI E battery cells are different from today's i3 battery cells, so further simulation and test work is mandatory.
- In the U.S. frequency regulation, V2G can lead to significant revenues of up to 1800 US $ per vehicle and year (@24 hour /7 days a week operation). Even stripped down to realistic availabilities, investment and operational cost token into account, real customer value seems to be a remaining outcome.

- Services as frequency regulation need a high availability of electric vehicles – ideally a 24 hour /7 days a week operation. If a customer cares to be plugged-in frequently at home and at work, 22 hours a day might be possible in theory. In such a case, the operational hours for at least the charging system and directly related E/E-systems inside the vehicle will rise up to more than 100.000 hours within a 12 years lifetime. Today's vehicles and specifications for most standard components are far below of that. Adoptions have to be brought on the way, as well as changes in the E/E system architectures to enable "stand alone operation" for the charging system including the direct communication links.

Figure 14: "Grid on Wheels": V2G demonstrator for frequency regulation

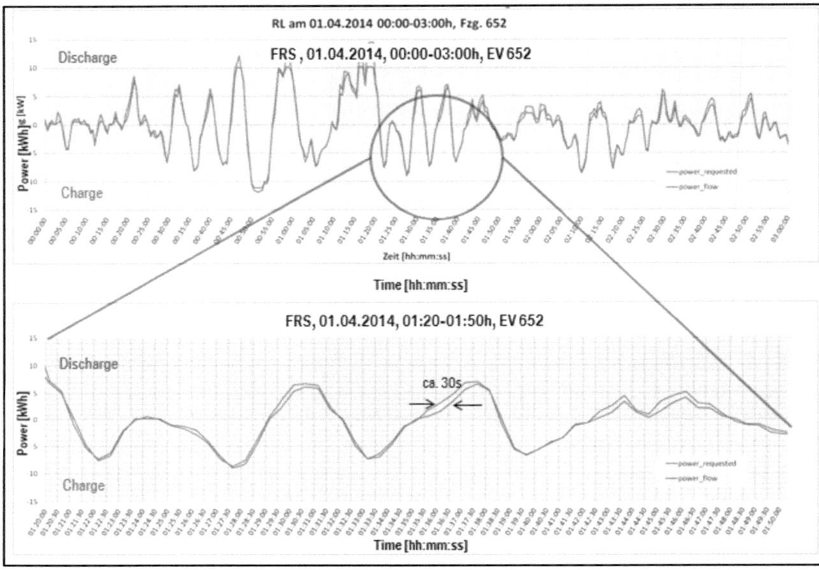

Figure 15: "Grid on Wheels": Typical Request (frequency regulation). V2G-enabled Mini E meets requirements.

For BMW the conclusion is: V2G and V2H as well may become a very valuable functionality to the customer and the energy economy in the future. But electric vehicles have to be prepared on it and a complete system approach is required to bring it to a success. Most important – beside of all the technical challenges – is the development of a sustainable business model and the proof of a positive business case for all stakeholders involved.

The shown two projects represent the bookend with the more ambitious ancillary services. The third project "ChargeForward" is aiming the other bookend "simple and viable with the existing BMW i3". In a cooperation with the Californian utility Pacific Gas & Electric (PG&E), since July 2015 the i3's of 100 BMW customers, who are living in the San Francisco Bay area, are member of a pool in combination with a stationary storage located at the BMW Tec office at Mountain view, CA. This pool is participating in the Demand Response and is activated usually once a day reducing grid load by 100kW as requested by PG&E. Customers have been incentivized with 1.000$ upfront at the beginning of their participation as a compensation for necessary installations like a separate meter for the charging location. They also allowed BMW to take over charge control and to interrupt charging events for at least one hour a day by the BMW "Connected Drive" telematics link.

The business model is simple: For every day participation in an occurring D/R-event customers receive 1$ extra, which may sum up to more than 300$ revenue per year for the customer. The money for that derives from 3.000$ payments for the frequent D/R delivery of 100kW. For possible future serial energy services, offered to electric vehicle customers, such business models have to be proofed in terms of a reliable pool control system even for far higher numbers of electric vehicles and a sustainable business case.

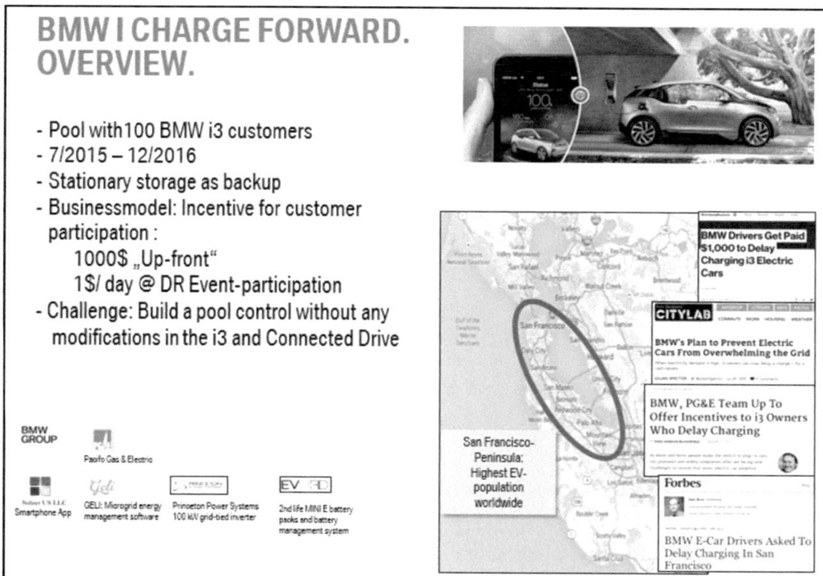

Figure 16: "ChargeForward": Demand Response demonstrator and pilot

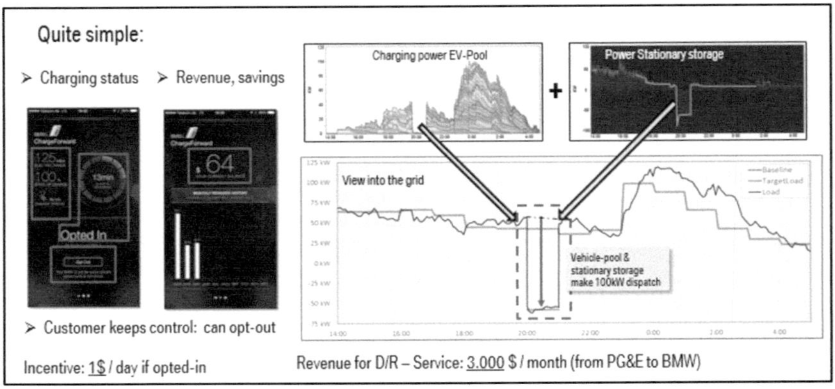

Figure 17: BMW i ChargeForward, interaction and effects

7 Regulatory framework

The existing legislation for the energy economy in most countries is designed for the existing system of power generation, distribution and consumption: generation is done by huge central power plants, feeding-in at the highest voltage levels, distribution happens vertically top-down. The smaller consumers as households, are charged directly according to the energy consumed in kilowatt-hours [kWh], bigger consumers above 100.000kWh in Germany, are charged by a peak-power rate in combination with the energy consumed. That's standard in most markets, wherein also the energy consumption of an electric vehicle gets counted.

For the further way of the energy transition which will bring a fundamental change to decentralized renewable power plants, The German government is preparing several new laws as the 'law on digitalization of the energy transition' (Gesetz zur Digitalisierung der Energiewende) and "electricity market 2.0" (Strommarkt 2.0).

If electric vehicles are getting more and more "grid friendly" as shown in this contribution, approaches for evaluating the value out of this have to be discussed more intensively right now, as the above laws are validated this year.

Together with the German automakers within the German Association of the Automotive Industry (VDA) and the national platform E-Mobility (NPE), several proposals are in discussion:

- Smart meter / Smart meter gateway regulations respect the communication channels between an electric vehicle and OEM backend systems
- Incentives for „grid-friendly“ charging infrastructure and intelligent connected devices (as electric vehicles, home storage, photovoltaic systems, heat pumps, home energy management systems)
- Reduced grid fees (according to German §14a EnWG, modified to E-Mobility constraints and requirements)
- Creating a legislation for the marketing of flexibility, even in small amounts typical for electric vehicles and feed-in from electric vehicles (V2G) by the further design of the new "electricity market 2.0" in Germany
- Easier access to electricity and ancillary markets
  - Reduction of minimum power limits for offers in the ancillary markets, smaller timing slot
  - Prequalification of standardized systems for the ancillary services according to the type approval for vehicles
  - Regulatory framework for a better sector coupling between traffic and energy

8 Conclusion

A proper grid integration of electric vehicles will support a "win-win"-situation for energy transition and E-Mobility as well:

The more electric vehicles, the more flexibility and storage capacity will be available for grid stability even under high amounts of renewable sources. The more renewables in the grid, the cleaner mobility will be.

- Therefore, a reliable, safe and cost effective communication between electric vehicles and backend systems will be mandatory – with respect to OEM liability responsibilities!
- Electric vehicle grid integration has to be in line with customers interests.
- Given markets and regulatory conditions assumed – BMW Group will develop electric vehicles and systems to support grid integration issues.

Sources

Bundesministerium für Wirtschaft, Zahlen und Fakten Energiedaten (Brutto-Stromvbrauch), 02/2013
DEW: installed power plant capacity 2014
Auer, Deutsche Bank (DB) Research: Moderne Stromspeicher, Unverzichtbare Bausteine der Energiewende, Jan 2012.

PREMIUM –Project (Plug-In-, Range-Extender- und Elektrofahrzeuge unter realen Mobilitäts- Umständen: Infrastruktur, Umweltbedingungen und Marktakzeptanz), BMW AG et alt.

„Merit Order Energiespeicher 2030", Research project, Forschungsstelle für Energiewirtschaft e.V. et alt.

„Challenges and results in the BMUB research project Controlled Charging V3.0"; Michael Westerburg et alt.; EWE AG

„Smart charging in daily routine – expectations, experiences, and preferences of potential users"; Franziska Schmalfuß, et alt; TU Chemnitz

"Local Energy Systems for Electric Mobility", Dr. Willibald Prestl, Bernhard Brendle, Dr. Michael Beer; BMW AG, Munich

Herausforderungen und Ergebnisse im BMUB Förderprojekt „Gesteuertes Laden V3.0“

(Challenges and results in the BMUB research project ‘Controlled Charging V3.0’)

Michael Westerburg, B. Jünemann, R. Drexler, EWE AG

Abstract

Der Energiemarkt wird sich grundlegend verändern und künftig zunehmend durch ein Zusammenspiel fluktuierender Erzeugungs- und dezentraler Speichereinheiten sowie traditionellen, konventionellen Erzeugern bestimmt. Diese Veränderungen erfordern eine weitere Flexibilisierung von Anlagen in Form von individualisierten Nachfrageprofilen, einer Synchronisierung der Volatilitäten und einer intelligenten Vernetzung der Komponenten im Rahmen eines Smart Grids. In diesem Kontext werden auch die Batterien von Elektrofahrzeugen weiter an Bedeutung gewinnen. Für die Steuerung der Stromflüsse bedarf es allerdings einer IKT-basierten Einbindung in das Smart Grid.

In dem vom Bundesministerium für Umwelt, Naturschutz, Bau und Reaktorsicherheit (BMUB) geförderten Verbundprojekt „Gesteuertes Laden V3.0" verfolgte das Konsortium das Ziel, eine Energie-Dienstleistung für Elektrofahrzeugbetreiber zu entwickeln. Ein intelligentes Energiemanagement, welches die Erfordernisse des Fahrzeugnutzers, der Energieerzeugung und die infrastrukturellen Rahmenbedingungen berücksichtigt, fand weitgehend automatisch ein technisches und wirtschaftliches Optimum für das Laden von mobilen Energiespeichern. Generiert wurde der Kundennutzen durch dessen Beteiligung am Erfolg der Vermarktung von Regelleistung.

Innerhalb des Förderprojektes lagen die wesentlichen Arbeitsschwerpunkte in der Bestimmung des Vermarktungsszenarios, in der Bereitstellung einer leistungsfähigen IKT und eines dezentralen Energiemanagementsystems, in der Erschließung der Speicherkapazität von Elektro-Fahrzeugbatterien und in der Steuerbarkeit der Ladevorgänge. Hier wurde eine Lösung erarbeitet, die die Erzeugung und Vermarktung von Regelleistung aus großen Elektro-Fahrzeugflotten ermöglichen kann.

Themenkreise

- Energiewirtschaft im Zeichen der Energiewende
- Märkte
- Energiezufuhr ins Fahrzeug
- Gesamtsystem

Kurzangabe über Neuigkeitswert der Arbeit

- Potenzialanalyse für die energiewirtschaftliche Vermarktung
- Berechnung von notwendigen Poolgrößen
- Validiertes Gesamtsystem

Das Forschungsprojekt "INEES" – Intelligente Netzanbindung von Elektrofahrzeugen zur Erbringung von Systemdienstleistungen

(The 'INEES' research project – intelligent grid integration of electric vehicles to provide system services)

Hannes Haupt, Dr. Georg Bäuml, Mathias Kammerlocher, Henning Nannen, Dr. Gunnar Bärwaldt

1 Einleitung

Das Konzept der Bundesregierung Deutschland für eine umweltschonende, zuverlässige und bezahlbare Energieversorgung definiert hohe Ziele für den Ausbau erneuerbarer Energieträger. Daher erlebt insbesondere die elektrische Energieversorgung aktuell einen fundamentalen Wandel. Immer mehr dezentrale, flexible Einheiten erzeugen Strom und sind dabei von fluktuierenden Faktoren wie der Sonneneinstrahlung oder dem Windaufkommen abhängig. Dadurch entstehen neben den bisherigen, durch schwankenden Stromverbrauch bedingten Fluktuationen zusätzliche kurzfristige Schwankungen im Stromnetz, die zur Erhaltung einer stabilen Netzführung ausgeglichen werden müssen. Wegen des Wegfalls konventioneller Kraftwerke, die bisher diesen Ausgleich vorgenommen haben, müssen hier neue Lösungen gefunden werden.

Gleichzeitig gewinnen im Verkehrssektor elektrisch angetriebene Fahrzeuge zunehmend an Bedeutung. Dabei stellt sich die Frage, wie diese Fahrzeuge möglichst vorteilhaft mit dem Stromnetz verbunden werden können.

Das Stromnetz benötigt den Ausgleich von Schwankungen auf unterschiedlichen Zeitskalen: Vom saisonalen Ausgleich zwischen Sommer und Winter bis hin zum Bereich von Sekunden für die Stabilisierung bei kurzfristigen Prognoseabweichungen. Die technischen Eigenschaften von Fahrzeugbatterien – hohe Leistung im Verhältnis zum Energieinhalt, schnelle Reaktion, Alterung bei Energiedurchsatz – legen nahe, gerade den letztgenannten, kurzfristigen Ausgleich genauer zu untersuchen, der in Form von Regelleistung erbracht wird. Denn dieser erfordert schnelle Reaktion bei hoher Leistung und vergleichsweise geringen Energieflüssen.

Ziel des Forschungsprojektes „INEES“ (Intelligente Netzanbindung von Elektrofahrzeugen zur Erbringung von Systemdienstleistungen) war es daher, die Möglichkeiten zu erforschen, aus einem Pool von Elektrofahrzeugen Regelleistung zu erbringen, genauer gesagt: Sekundärregelleistung, also der Ausgleich von Schwankungen im Bereich von 30 Sekunden bis 15 Minuten.

2 Das Gesamtsystem

Um das Ziel zu erreichen wurde basierend auf dem aktuellen Stand der Technik im Jahr 2014 ein Gesamtsystem aufgebaut, das die technischen Anforderungen an die Erbringung von Regelleistung aus Elektrofahrzeugen aufgreift und soweit möglich erfüllt.

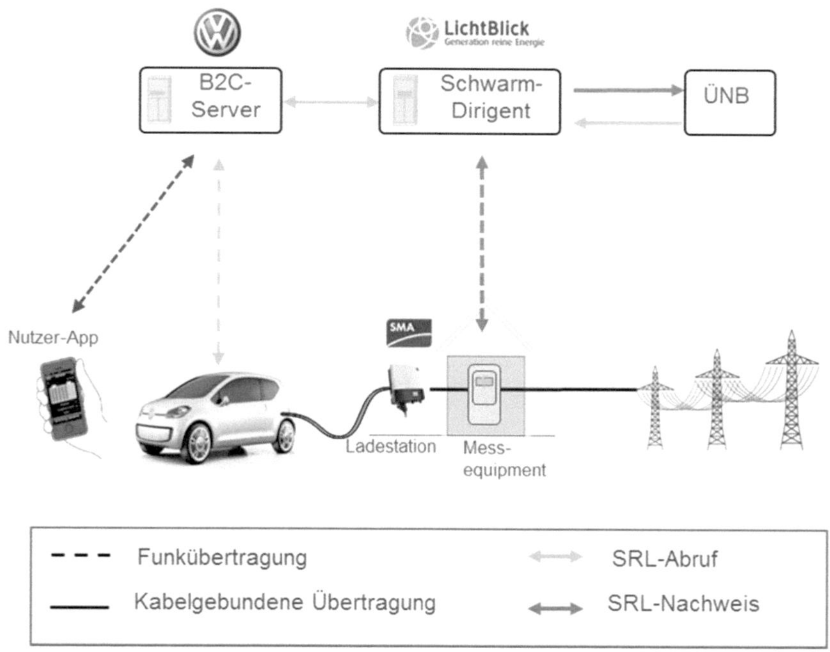

Abbildung 1: Das INEES-Gesamtsystem

Dazu wurde von der SMA Solar Technology AG eine als Experimentiersystem konzipierte, bidirektionale DC-Ladestation (10 kW) entwickelt und eine Kleinserie von 40 Geräten aufgebaut. Weiterhin wurden von der Volkswagen AG 20 e-up!s mit einer bidirektionalen Ladefunktion ausgestattet und eine Kommunikationsanbindung zwischen Ladesteuerung und Volkswagen-Backend implementiert. Des Weiteren wurde eine Mobiltelefon-App als Nutzerschnittstelle entwickelt. Als Schnittstelle zwischen Elektrofahrzeugen und Energiemärkten wurde von der LichtBlick SE der SchwarmDirigent® um die Integration der Fahrzeuge erweitert. Abbildung 1 zeigt das aufgebaute Gesamtsystem.

Darüber hinaus hat das Fraunhofer IWES als Projektpartner die wissenschaftliche Begleitforschung zu Themen Belastung der Verteilnetze und volkswirtschaftlicher Nutzen der Erbringung von Regelleistung aus Elektrofahrzeugen übernommen.

3 Technische Machbarkeit

Als Basisfahrzeuge wurden 22 Volkswagen-e-up!-Serienfahrzeuge mit Combined Charging System[1]-Ladedose verwendet.

Abbildung 2: Volkswagen e-up!

Um die INEES-Funktionalitäten zu realisieren, wurden zusätzliche Forschungskomponenten in den Fahrzeugen verbaut. Dies umfasste eine Car-Telematic-Box (CaTe) inklusive GPS/GSM-Antenne, die mithilfe eines zusätzlichen Kabelstranges an einzelne Fahrzeugdatennetze angeschlossen und über eine proprietäre Leitung mit dem Lademanager verbunden wurde. Zu Analysezwecken wurde ebenfalls in der Kofferraummulde ein Datenlogger inkl. GPS/GSM-Antenne verbaut.

1 *Combined Charging System*; Schnellladestandard der europäischen Automobilhersteller

Die Erbringung von Sekundärregelleistung setzt eine dynamische Regelung des Gesamtsystems voraus. So muss beispielsweise einem vom Übertragungsnetzbetreiber (ÜNB) vorgegebenen Sollsignal nach spätestens 30 Sekunden mit einer Leistungsänderung des Pools gefolgt werden. Die volle abrufbare Leistung muss nach spätestens fünf Minuten erbracht werden. Neben dieser Dynamik ist dem ÜNB ein wiederkehrendes, systematisches Regelverhalten wichtig.

Eine besondere Herausforderung im Projekt INEES waren die Abbildung von im Ruhezustand befindlichen Fahrzeugen und das Reaktivieren über die Betriebssteuerung des Poolmanagers. Angeschlossene Fahrzeuge wechseln nach einer kurzen Zeitspanne der Inaktivität selbstständig in einen Ruhezustand und trennen sich kommunikativ und elektrisch von der Ladestation. Genauso trennt sich die Ladestation elektrisch vom Stromnetz und wechselt in einen Ruhezustand. Eine erneute Netzkopplung kann bis zu 300 Sekunden dauern, da verschiedene Prüf- und Sicherheitsphasen vor der Wiederverbindung mit dem Stromnetz durchgeführt werden müssen (Vgl. z.B. Technische Anschlussbedingungen Niederspannung). Für die Erbringung von Sekundärregelleistung ist eine Reaktion unter 30 Sekunden erforderlich, sodass im Zielsystem eine deutlich schnellere Wiederaufnahme der Leistungserbringung nötig würde. Mögliche Lösungen wären, die Trennung der Ladestation und des Fahrzeuges vom Stromnetz zu unterbinden oder die Anforderungen zur Leistungserbringung unter Berücksichtigung dieser Rahmenbedingungen zu überarbeiten.

Zusammenfassend wurde im Projekt nachgewiesen, dass die Erbringung von Regelleistung mit einem Elektrofahrzeugpool technisch möglich ist. Dies konnte im Flottenversuch demonstriert werden, wobei einige Fragestellungen auf Grund der begrenzten Fahrzeuganzahl nur mit Hilfe unterstützender Simulationen untersucht werden konnten.

Auf Grund der hohen Dynamik der Entwicklung des Energiesystems ist weiterhin zu evaluieren, inwieweit zukünftige Anforderungen mit diesem Konzept zu vereinbaren sind.

4 Nutzerakzeptanz

Eine wesentliche Voraussetzung für den erfolgreichen Einsatz von Elektrofahrzeugen zur Erbringung von Systemdienstleistungen ist die Partizipation der Nutzer des Systems. Dabei ist insbesondere der Wunsch nach einem flexiblen und spontanen Einsatz der Elektrofahrzeuge zu berücksichtigen. Demgegenüber steht Bedarf für eine verlässliche Planbarkeit der für die Systemunterstützung zur Verfügung stehenden Kapazität. Eine Möglichkeit diese Ziele in Einklang zu bringen ist in Abbildung 3 dargestellt.

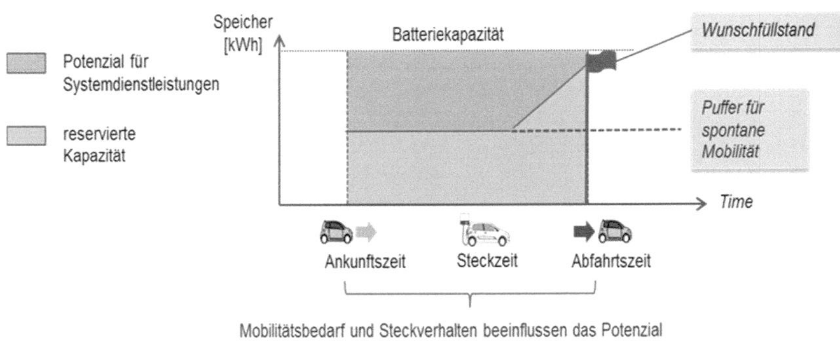

Abbildung 3: Grundeinstellungen zur Freigabe von Kapazität

Der Nutzer reserviert einen frei wählbaren Anteil seiner Kapazität als Puffer für spontane Mobilität („reservierte Kapazität"). Dieser wird jederzeit garantiert vorgehalten. Der darüber hinaus gehende Teil seiner Kapazität kann zum Zweck der Bereitstellung von Systemdienstleistungen für Lade- und Entladevorgänge genutzt werden. Für Fahrstrecken, die über den Bedarf für spontane Mobilität hinausgehen, kann der Nutzer einen Wunschfüllstand zu einem beliebigen Abfahrtszeitpunkt reservieren. Dieser Füllstand wird wiederum bis zur tatsächlichen Abfahrt oder aktiven Freigabe reserviert. Im Projekt INEES konnten diese Einstellungen sowohl im Fahrzeug selbst, als auch über eine Smartphone-App vorgenommen werden. Abbildung 4 zeigt die Umsetzung in der Smartphone App.

Abbildung 4: Umsetzung der Nutzereinstellungen in einer App

4.1 Verfügbarkeit der Fahrzeuge aus Mobilitätssicht

Damit die Elektrofahrzeuge Systemdienstleistungen für das Stromnetz erbringen können, müssen sie mit einer Ladestation verbunden werden. Um das Potenzial bestmöglich auszuschöpfen sollte es sich dabei um eine bidirektionale Ladestation handeln. Im Rahmen des Projektes INEES wurde jedem Teilnehmer eine solche Ladestation am Standort „Zuhause“ installiert. Die Verfügbarkeit an dieser Ladestation hängt stark mit dem Mobilitätsverhalten zusammen. Abbildung 5 zeigt daher die Verteilung des Mobilitätsverhaltens der Teilnehmer am Projekt INEES. Dabei wurde einigen Pendlern zusätzlich zur Ladestation am Standort „Zuhause“ eine weitere Ladestation an Ihrem Arbeitsort installiert. Dadurch sollte untersucht werden, wie stark die Verfügbarkeit der Fahrzeuge durch eine solche Maßnahme gesteigert werden kann.

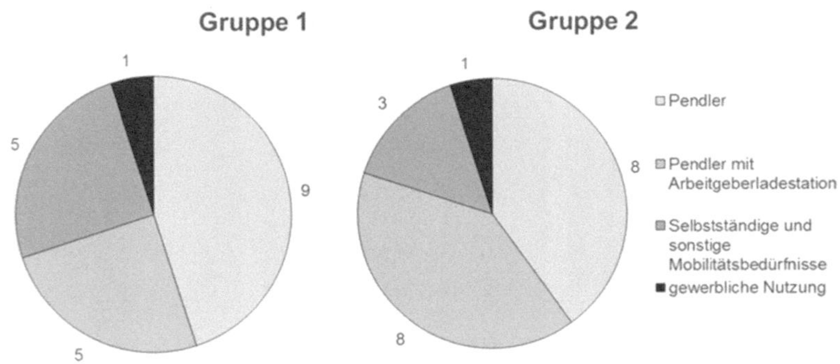

Abbildung 5: Mobilitätsprofile der Teilnehmer am Flottenversuch

Insgesamt ergab sich aus dem Mobilitätsverhalten eine sehr hohe Verfügbarkeit von Fahrzeugen an den Ladestationen und somit eine gute Eignung für die Bereitstellung von Systemdienstleistungen. Abbildung 6 zeigt, dass in der Regel zu jedem Zeitpunkt etwa 40 % der teilnehmenden Fahrzeuge mit einer INEES-Ladestation verbunden waren.

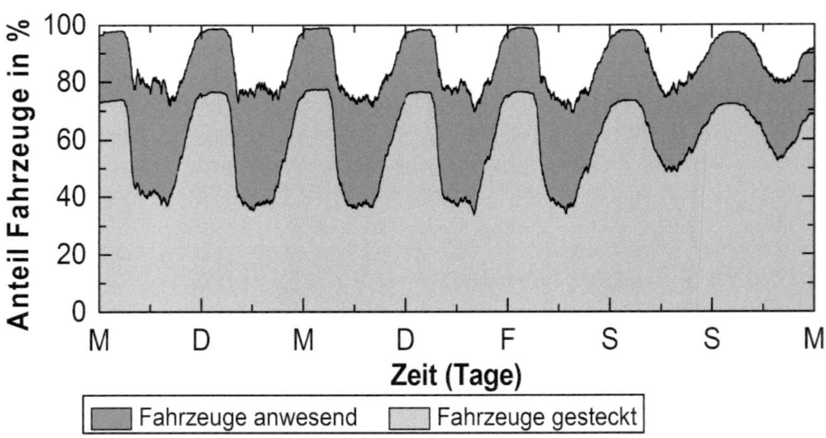

Abbildung 6: Verfügbarkeit der Fahrzeuge an einer INEES-Ladestation

4.2 Freigabe durch die Nutzer

Wie bereits beschrieben wurde, ist die Sicherstellung eines flexiblen und spontanen Fahrzeugeinsatzes ein wesentlicher Einflussfaktor auf die Akzeptanz der Nutzer zur Freigabe ihres Elektrofahrzeuges zur Unterstützung des Stromnetzes.

Im Mittel haben die Teilnehmer am Flottenversuch 34 % Ihrer Kapazität als Puffer für spontane Mobilität reserviert. Somit standen noch 66 % der Kapazität für die energiewirtschaftliche Nutzung zur Verfügung.

Abbildung 7 zeigt die Verteilung der durchschnittlichen Mindestfüllstand-Einstellungen.

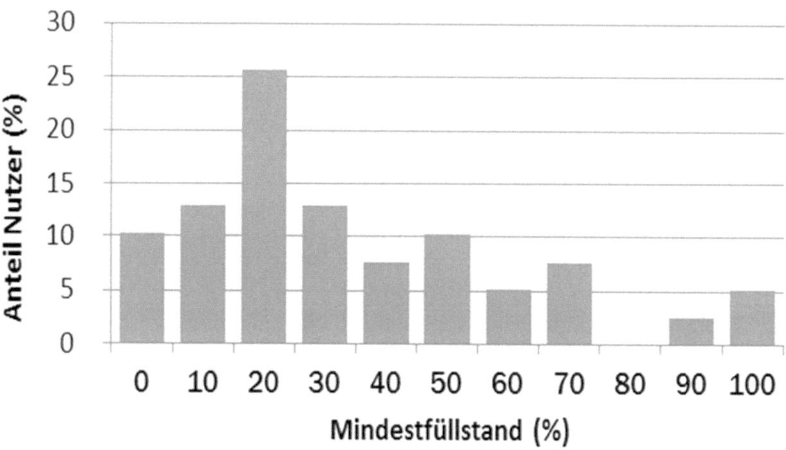

Abbildung 7: Verteilung der Einstellung „Mindestfüllstand"

Dabei wurde die Einstellung des Mindestfüllstands teilweise von technischen Problemen beim Ladevorgang der proprietären bidirektionalen Anbindung, welche zu Ladeabbrüchen führte, beeinflusst. Einige Teilnehmer berichteten, zur Sicherheit einen höheren Mindestfüllstand eingestellt zu haben als eigentlich von Ihnen für notwendig erachtet wurde. Sie würden diesen reduzieren, wenn die Technik einen zuverlässigen Serienstand habe. Darüber hinaus können sich die Nutzer laut eigenen Aussagen eine höhere Freigabe von Batteriekapazität vorstellen, wenn die Reichweite des e-Autos größer wäre.

4.3 Fazit Nutzerakzeptanz

Der Flottenversuch INEES zeigte, dass die Fahrzeuge sehr regelmäßig mit einer INEES-Ladestation verbunden werden und dass ein ausreichend hoher Anteil der Batteriekapazität der Unterstützung des Stromnetzes zur Verfügung gestellt wurde. Allerdings zeigte sich auch, dass die Freigabe sowohl unter den Nutzern als auch im Zeitverlauf stark schwankte.

Es wurden daher mehrere Befragungen durchgeführt, um zu erfahren, aus welchen Gründen sich die Nutzer für die Teilnahme am Flottenversuch und eine hohe oder niedrige Freigabe entschieden haben.

Als Hauptgrund für die Teilnahme am Flottenversuch und die Freigabe von Batteriekapazität gaben die Teilnehmer den Umweltschutz an, „Vorbild für nachfolgende Generationen sein", „Unterstützung des Forschungsprojekts" und die „Unterstützung der Energiewende".

Weiterhin wurden die für die Teilnahme notwendigen Komponenten den Teilnehmern unentgeltlich zur Verfügung gestellt. Lediglich die Fahrstromkosten mussten durch die Teilnehmer selbst getragen werden. Für die Freigabe von Batteriekapazität erhielten sie eine zusätzliche Gutschrift auf diese Kosten, deren Höhe sich nach Ihrer eingestellten Freigabe von Kapazität und ihrer Standzeit an einer INEES-Ladestation richtete.

Ab einem monatlichen Betrag über 10 €/Monat wurde von einigen Teilnehmern auch die Prämie als Grund für die Freigabe von Batteriekapazität genannt. Dies spiegelt sich auch in der Bewertung der Prämie durch die Teilnehmer in der Befragung wider. Während die Teilnehmer der ersten Gruppe die Prämie nahezu geschlossen als zu niedrig bewerteten und als hauptsächliche Motivation für die Teilnahme die Unterstützung des Umweltschutzes angaben, bewerteten die Teilnehmer der zweiten Gruppe eine höhere Prämie als durchaus interessanten Anreiz zur Teilnahme am Projekt.

Zusammenfassend kann festgehalten werden, dass die Teilnehmer am Flottenversuch ihre Fahrzeugbatterie mit großer Begeisterung für die Unterstützung des Stromnetzes freigegeben und dabei grundsätzlich keine Einschränkungen in ihrer alltäglichen Mobilität erlebt haben.

5 Ausblick

Bei der Weiterentwicklung der laderelevanten Funktionen von Elektrofahrzeugen steht der Kundennutzen im Mittelpunkt. Da 80 Prozent der Ladeereignisse an heimischen Ladeorten zu erwarten sind, werden erste Teilfunktionen, die Netzintegrationsaspekte aufgreifen, in diesem Kontext zu betrachten sein. Die Orientierung des Ladens an das Dargebot lokal erzeugter elektrischer Energie mit dem Ziel der Eigenverbrauchsmaximierung stellt dabei eine solche Teilfunktion dar. Für die Umsetzung dieser Funktion ist der Einsatz eines Energiemanagementsystems (EMS) zu erwarten, das sich in der Kundenumgebung befindet. Dieses hat zur Aufgabe, neben dem Energiebedürfnis des Fahrzeugs auch alle weiteren lokalen Energiebedürfnisse je nach Kundenwunsch hinsichtlich Zeitpunkt und Leistung z.B. an der lokalen Stromerzeugung auszurichten. Das Fahrzeug erhielte in diesem Fall monetäre Anreize vom EMS und antwortete darauf unter Berücksichtigung der Kundenvorgabe (Kostenminimal, ausschließlich EE ...) mit seinem Ladeplan. Das EMS lässt den Ladeplan in die weitere energetische Optimierung der Kundenumgebung einfließen.

Die im Projekt INEES untersuchte Funktion „Erbringung von Sekundärregelleistung" ist eine weitere Teilfunktion, die überregionale Schwankungen ausgleichen hilft. Beide Teilfunktionen gilt es zukünftig in geeigneter Weise zusammenzuführen, sodass für den Kunden sowohl lokal als auch überregional ein größtmöglicher Nutzen entstehen kann, ohne hierbei den Kundennutzen einzuschränken. Im Projekt INEES wurde ein proprietärer Kommunikationskanal genutzt, der eine kurze Latenzzeit der Signalkette ermöglichte, jedoch im Massenmarkt ungeeignet ist, eine lokale Optimierung effizient zu unterstützen. Eine Berücksichtigung lokaler Optimierungsziele war jedoch nicht Bestandteil des Projektes.

In einem künftigen Szenario könnte im Vergleich zu diesem proprietären Kommunikationskanal nicht das Fahrzeug der Empfänger von Signalen zur spontanen Ladeleistungsbeeinflussung sein, sondern das EMS in der Kundenumgebung. Das EMS könnte die Flexibilität aller intelligenten Verbraucher (wie z. B. des Elektrofahrzeugs) ermitteln und die gesamte Flexibilität des Haushaltes einer übergeordneten Instanz anbieten, die diese Flexibilität gegen Entgelt nutzen möchte. Der Abruf der Flexibilität könnte sich dann an den aktuellen Funktionsprinzipien der Regelleistung orientieren und eine Weiterentwicklung dieser darstellen. Dieser Ansatz wirft jedoch offene Fragestellungen hinsichtlich der energiewirtschaftlichen Abwicklung und technischen Umsetzung auf, die im Zusammenhang mit der Erbringung von Regelenergie zu beachten sind. Diese sind in weiteren Forschungsarbeiten zu untersuchen.

Danksagung

Wir danken dem Bundesministerium für Umwelt, Naturschutz, Bau und Reaktorsicherheit sowie dem Projektträger VDI/VDE Innovation und Technik GmbH für die Förderung des Projekts (FKZ 16EM1015). Die Verantwortung für den Inhalt der Veröffentlichung liegt bei den Autoren.

KEYNOTE LECTURE

Roles for regulators in electric Vehicle-Grid Integration

Noel Crisostomo, California Public Utilities Commission, USA

Abstract

California's principal energy agencies are currently establishing the policy and regulatory foundations for the widespread transformation of the transportation sector to one that is powered by decarbonized electricity. Executive orders and legislation established the basic frameworks needed to deploy vehicle technologies that serve the complementary goals to reduce transportation-related criteria air pollution in non-attainment air quality districts 80 % from current levels by 2030 and greenhouse gas emissions statewide 80 % below 1990 levels by 2050. In support of the California Air Resources Board (CARB) and California Energy Commission (CEC), the California Public Utilities Commission (CPUC) develops utility policies in four broad program categories, consistent with milestones set forth in orders and law:

- Coordinating the buildout of infrastructure to charge 1 million ZEV by 2020;
- Encouraging vehicle adoption such that 1.5 million ZEVs are driven on roads by 2025;
- Designing rates and incentives for low carbon fuels to halve petroleum use by 2030; and
- Utilizing Vehicle-Grid Integration technologies to use transportation energy as a resource that facilitates a 50 % renewable electricity system by 2030.

In response to the State's 2008 Climate Change Scoping Plan, which identified that efficient, zero-emission vehicles using low carbon fuels are 'needed as vital components' for the State to achieve its GHG emission reduction goals, Senate Bill 626 authorized the CPUC to establish rules to support the widespread deployment and use of plug-in hybrid and electric vehicles. Foundational conclusions during the CPUC's subsequent rulemaking – that providers of electric vehicle charging services are not subject to regulation as a public utility and that the investor-owned utilities would be prohibited from owning charging infrastructure – established the growth dynamics of the initial stages of the electric vehicle market. In 2013, CPUC continued to develop rules with a specific focus on plug-in electric vehicles to ensure that they efficiently integrate with the utility grid and access fair rates that encourage electrification, and to explore how the utilities could finance charging equipment and infrastructure. With the belief that privately-funded charging infrastructure installation would be scaled at a slower rate than needed to meet the State's goals, CPUC removed the prohibition on utility infrastructure ownership. In total, three large investor-owned electric utilities proposed to build 60,600 electric vehicle service equipment connectors at a cost of approximately $ 1.1 Billion.

Through a vigorously-contended public process that included consumer, environmental, charging industry, and automotive industry input, the CPUC established the scope and nature of utility infrastructure programs, from a market design and technological perspective, to ensure just and reasonable expenditures of utility customer funds. These first-of-a-kind utility vehicle charging infrastructure programs from CPUC will, in support of complementary initiatives at CARB and CEC, inform how the State will scale up to levels needed for 'widespread transportation electrification'.

In 2015, Senate Bill 350 codified that widespread electrification of all vehicles, vessels, rail, boats, and other equipment would be a 'principal goal' of the utilities, alongside core utility functions of resource planning and procurement. Commensurate to the scale of adoption required and in consideration of both the commercially-available products and advanced technologies being supported by State research and development funds, it is imperative to use electric vehicles to assist in grid management, integrate renewable energy generation, and reduce fuel costs for vehicle drivers who charge in a manner consistent with electrical grid conditions, also known as 'Vehicle-Grid Integration' (VGI).

VGI poses several opportunities and challenges in its proposition to use customer transportation assets that are flexible, geographically dispersed, mobile, require large-scale deployments of enabling infrastructure, and are undergoing simultaneous technological change and regulatory development. United around common objectives, the State agencies' tangential jurisdictional authorities of air resources, energy policy, and utility regulation provide the breadth and depth of oversight necessary to facilitate VGI. Key to this dynamic is the State's ability to catalyze the evolution of automotive original equipment manufacturers' mobility strategies and to foster private enterprise in charging networks and services in a fashion that complements public and utility customer investments.

The State is on the verge of decisions, rule reformation, and program deployments that will implicate the achievement of its clean energy and environmental objectives. Vehicle-Grid Integration policy developed at the regulatory agencies will establish how the automotive, transportation, and electricity sectors converge through the design of energy market tariffs and the utility accommodation of mobile and distributed resources in grid planning, and modifications to mandates affecting vehicle sales and mobile air pollution. This paper maps these and other extant electric transportation initiatives relative to one another; proposes common operating principles and coordinating tasks that will assist in the coordination of State and similar United States Department of Energy grid modernization efforts; and identifies key contingencies between regulations that affect future utility customer programs, vehicle designs, service equipment capabilities, and mobility businesses.

LOCAL ENERGY SYSTEMS

Local energy systems for electric mobility

Dr. Willibald Prestl, Bernhard Brendle, Dr. Michael Beer

BMW AG, Munich

1 Charging solutions as interface to energy systems

The transition to sustainable energies does not only include renewable energy generation, but also efficient energy storage and a flexible operation of power consumers. Electric mobility offers great potential for such optimizations.

Electrical Powertrains allow all primary renewable energy sources – solar power, wind energy and water power – to be used for mobility purposes. Combining electric mobility and the potentially large flexibility of its charging processes with local energy supplies and stationary storage batteries has the potential to bring cost optimizations and reduce negative impacts on the electricity network. To exploit these potentials it is essential to implement intelligent charging solutions, which require not only the vehicle and its charging infrastructure, but also a sophisticated and integrated energy management system. The system must be compatible with circumstances of the vehicle operator and allow an optimization of the energy usage in a way that is beneficial for the user and the electricity network.

BMW researches and develops intelligent charging systems in various projects in order to make local energy generation available for the charging of electric vehicles (EVs). These are the functionalities that will form an interface between electric mobility and sustainable energy systems.

2 Customer needs and requirements of intelligent charging solutions

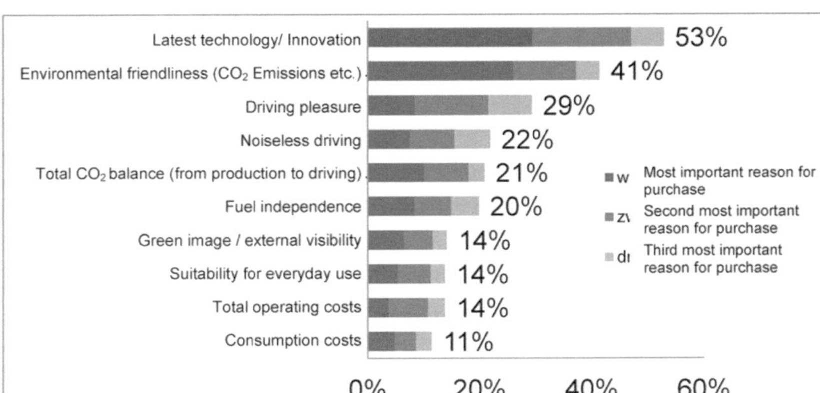

Figure 1: Reasons for purchasing EVs

Intelligent charging processes that exceed the capabilities of standard (immediate) charging can only be launched on the market if they satisfy crucial customer needs. Figure 1 shows the results of a survey of 284 BMW i3 users from the project PREMIUM[1].

According to these results, EV users are very interested in innovative products, but the environmental impact of local emissions and of the energy supply are also important factors:

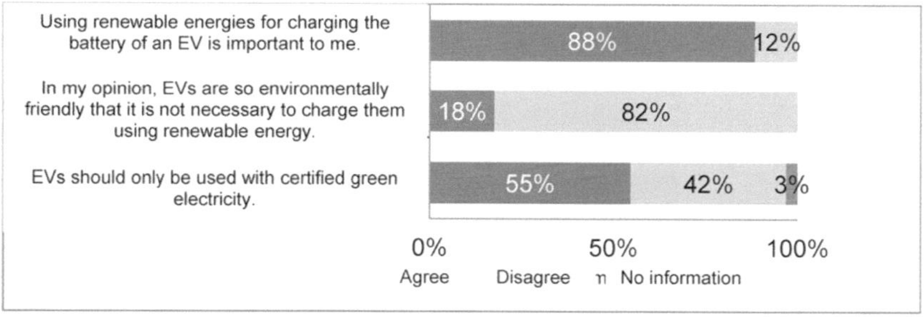

Figure 2: Statements about environmentally friendly charging of EVs

The project PREMIUM was also used to collect data about the usage and charging behavior of the vehicle as well as the users' charging circumstances:

The private usage conditions can be described as follows:

- 95 % of customers use charging options at home.
- 80 % of customers live in single-family or terrace houses, 74 % have a private garage or parking space. Only 5 % have no option to charge at home.
- The average plug-in time is above 9 h. This offers great flexibility for an optimized planning of charging cycles.
- 71 % use green electricity, 45 % use photovoltaics for domestic energy generation

Therefore private customers are an excellent starting point for offering charging solutions and services that connect electric mobility to sustainable energy supplies.

Only 20 % of users charge their EV at their workplace, even though the majority of journeys serve for commuting or business. This shows that apparently there still exist barriers to investments in charging infrastructure at workplaces. The requirements for fleet charging solutions (daily commuter fleets and fleet operation from the operator's location) that are different from private usage are listed in table 1.

1 *www.premium.uni-passau.de; Center for Market Research Passau University.*

Table 1: Summary of customer requirements for charging solutions

Private charging	Fleet charging
• Simplicity of charging process and customer interaction • Sustainable energy supply, usage of local generation • Optimal integration into the local energy system, e.g. load limiting, generation, storage battery	

• Minimization of electricity cost (operating price)
• Experience of sustainable mobility
	• Avoiding/limiting load peaks
	  • Secure reaching of charging targets
	  • Minimization of electricity cost (operating + peak demand charge)
	  • Avoid installation efforts
	• Compatibility with relevant standards
	• Functionality independent of car manufacturer
	• Administration-, reporting and billing tools

3 Concept and energy management for fleet charging solutions

Energy management is crucial for charging electric or plug-in-hybrid fleets, especially because few vehicles can suffice to reach the power limit of an electric installation. This requires a variable limitation of the charging power for fleet charging.

In commercial installations there is an increased risk that an unfavorable interaction of power peaks of the fleet charging infrastructure and other electrical loads can lead to increased electricity costs. In the case of obligatory power metering, the maximum yearly power peak is a decisive factor to determine the peak demand charges. Together with the operating price these add up to the total energy costs. Avoiding costs caused by power peaks is an important requirement for the energy management system.

In addition to limiting the charging power to meet technical or monetary requirements, an important aspect is to consider how vehicle fleets can be optimally charged using renewable energies available e.g. at their company's location. In this case, the energy management system must be able to vary the fleet's charging process according to the availability of locally generated power. This does not only bring cost advantages but also leads to a sustainable mobility using renewable energies and helps to avoid undesired peaks in the grid feed.

Furthermore, the German government is discussing restrictions for the grid connection of charging infrastructures as a means to avoid network imbalances. The currently available vehicles themselves do not cause improper grid imbalances. Charging multiple electric or plug-in-hybrid vehicles with single-phase chargers at the same grid connection point on the same phase however can cause such imbalances. These current discussions lead to further requirements for the load and energy management of fleet charging solutions.

3.1 State of the art

Currently available solutions mainly aim to not exceed the maximum power limit of a connection point or an electric installation. These commercial solutions are for example capable of distributing the available system power equally to multiple vehicles or charging processes. Another approach is to distribute the available power based on strategies such as the 'First in – First out' principle.

All these strategies are reactive and without advance preplanning. In the case that the power limit might be exceeded by a charging process, the existing fleet charging solutions react and reduce the charging power of one or many vehicles.

For this, fleet charging solutions are equipped with local control units that are usually installed in the sub-distribution part of the system, which is connected to all charging points. This connection is used to control the charging power available at each charging point.

The communication between vehicle and charging point is based on the IEC 61851 standard and the power control of charging processes is controlled by the so-called PWM-signal via the pilot contact. However, there is no uniform standard for the communication between the control unit and the charging points, and even proprietary solutions are partially being used. During the charging process according to IEC61851 the duty cycle of the PWM-signal sends information about the maximum allowed charging current to the vehicle charger. In case of a load management action, current fleet charging solutions vary this limit during the charging process.

3.2 Infrastructure-based charging point control

None of the currently available systems is capable of pre-planning charging processes based on departure times and energy demands and thus terminate or control charging processes based on these criteria.

Using the experiences gained from the research project ‚*Energy self-sufficient electric mobility in a smart-micro-grid from a single-family house to a car park*' [2], the developments of the BMW fleet charging solutions are based on a standardized approach,

2 http://www.elektromobilitaet-verbindet.de/projekte/energieautarke-elektromobilitaet.html

which allows a transmission of charging plans with variable charging profiles via OCPP. The power control and the execution of the plans can be performed using the PWM-signal. OCPP is a communication protocol of the Open Charge Alliance, which aims to promote a worldwide development, implementation and adherence of communication protocols for charging infrastructures and the resulting necessary standards. Starting from the version OCPP 1.6, the protocol is capable of communicating a charging profile from a central system to a charge point.[3]

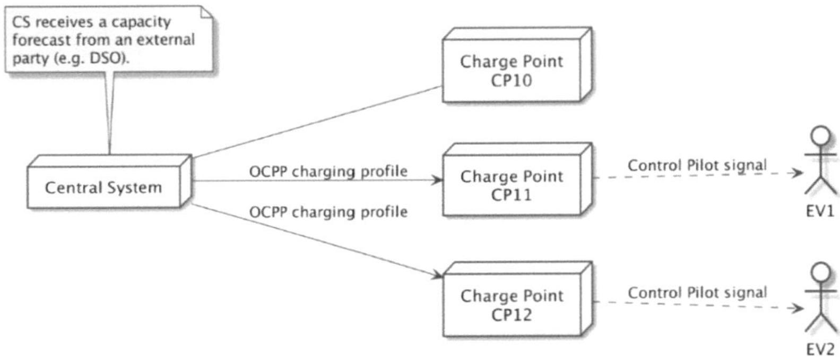

Figure 3: Smart Charging architecture of OCPP 1.6[4]

3.3 Elements of an energy management for fleet charging solutions

A modern load and energy management of fleet charging solutions can be described in the form of two blocks. The first one is a load management block that operates only in a reactive way and without proactive energy planning. The other one is a predictive module that is capable of creating optimally coordinated charging plans when information about planned departure times and energy demands are available.

Load management block

If no information about user and vehicle is available, this block must perform phase-specific load management and limit the phase imbalances and assure the limitation of the overall power peak. Single-phase vehicles which charge on the same grid phase

3 http://www.openchargealliance.org/

4 Open Charge Alliance; Open Charge Point Protocol 1.6; Interface description between Charge Point and Central System; Document Version 1.6; 2015-10-08

are thereby directly competing for charging power. Three-phase vehicles only need to be considered if there is a maximum overall power limitation, as they do not contribute to phase imbalances.

If no prioritization of vehicles is performed, e.g. based on the plug-in time, all vehicles in direct competition will be charged with the same power, considering the upper power limit and the network asymmetries at each point of time. The load management block needs to check periodically whether the requirements and the actual powers drawn by the vehicles have changed. This allows for example to free the power of a charging point once the vehicle at this charging point is fully charged.

Predictive charge planning

Resulting from the outcomes of the project *‚e-MOBILie'*, a predictive energy management must be capable of generating charging plans that consider decentralized energy generators, such as the expected generation power of a PV installation. Furthermore it is necessary to adhere to tariff boundaries and technically determined connection limits.

The optimization targets which are used to generate predictive charging plans cannot be generally described, but the requirements for the energy management are the deciding factors here. These can be motivated by technical, monetary or sustainability aspects.

One strategy might be to maximize the self-usage of PV power and at the same time reduce peaks in the grid supply. Such an energy management strategy was developed using simulation models. An abstract of the results is shown in figure 4. For a configuration of a vehicle fleet with a PV installation the optimization strategy is based on a prioritization. In the first instance, the charging energy is covered by self-generation, shown in figure 4 (left).

A further step plans the power curve of the grid supply, considering the supply limitations in order to smooth out the draw on the mains. Using multiple of such limitation curves allows an avoiding of power peaks and adjusting a technical connection limit. The red curve in figure 4 (left) shows the charging power curve of a fleet without load and energy management, while the charging curve of a fleet that considers PV generation (brown) is shown in blue. The right figure also considers a technical connection limit.

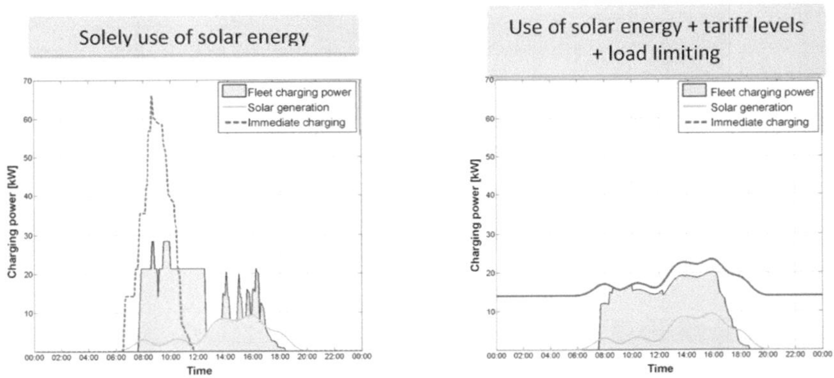

Figure 4: Optimization of the charging processes of a vehicle fleet

Similar to load management, periodic checks are necessary to determine whether the framework conditions have changed. In this case a rescheduling of the charging processes is necessary. In reality, this means that for example a vehicles leaves earlier than expected or increases its charging power because preconditioning was started. All plans need to be generated such that at every point of time the requirements for limitations and maximum network imbalances are fulfilled.

3.4 User benefits

The benefits of using an energy management system for fleet charging has two important aspects: On the one hand, it offers clear benefits for the operator of a charging infrastructure, as he saves cost for the installation and extension of electric infrastructures. By the means of load and energy management it is no longer necessary to design an infrastructure for a simultaneity factor of 1. Such a solution furthermore allows a combination of electric mobility with other power generators such as PV-plants or storage batteries and possibly reduces the peak demand charge.

On the other hand, an intelligent fleet charging solution offers benefits for the network operator. In this way, network imbalances can be reduced and power peaks or peaks of grid feed-in of renewable generation can be diminished.

4 Concept and energy management for home charging

4.1 Operating principle and system architecture

The major customer requirements for home charging solutions are a minimization of cost and a maximization of the use of self-generated energy from domestic photovoltaic (PV) installations. An energy management system that optimizes charging processes and energy flows in the house simultaneously can automatically contribute to fulfilling these requirements. This principle is based on flexible energy consuming appliances, which are planned on a daily basis using an appropriate optimization algorithm. It takes account of the existing base load and other fixed loads and achieves the optimization targets e.g. by making use of predicted solar energy generation.

The process of charging EVs, for example, can be split in several sub-blocks, which can be timed as needed and adapted to a power curve. The main constraint for the optimization is hereby that the vehicle is charged to the level required by the user at the planned departure time.

Such a system was developed in the electric mobility showcase project '*e-MOBILie: Energy self-sufficient electric mobility in a smart-micro-grid*' [2], a collaboration of BMW AG, SMA solar technology AG and the Technical University (TU) Munich. It was tested in a 1-year pilot run.

A system architecture using the future charging communication standard ISO15118 was implemented here. This vastly increases the charging control compared to a basic charging functionality defined by IEC61851 as explained in chapter 3. Figure 5 shows the system configuration and the process of generating a charging plan (❶-❹).

The vehicle plans and controls the charging process triggered by an external incentive system. It can respond to the incentives while optimally servicing its own system parameters such as battery, charging system and relevant consumer devices.

The incentive system is a data structure that can contain abstract cost tables as well as power limitations. Its generic form allows transmitting data from various high-level optimization systems in a uniform way, securing a high degree of universality of the interface.

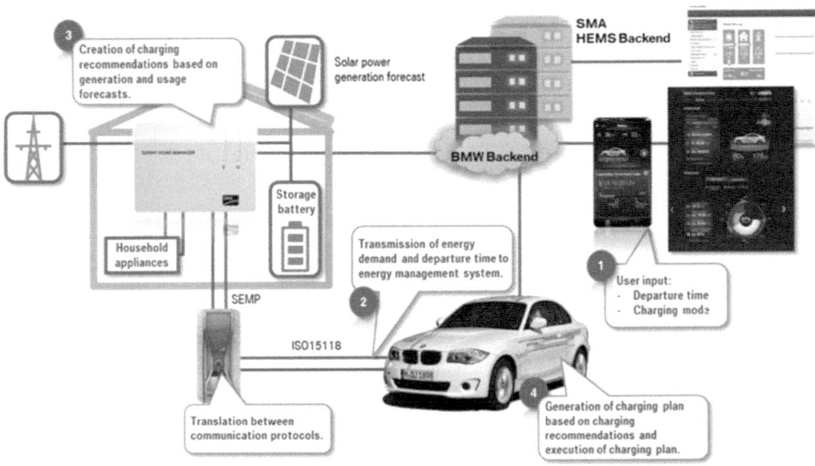

Figure 5: Structure and functionality of an integrated energy management system that connects vehicle charging to a home energy system

4.2 Customer benefits

A simulation model was used to analyze different types of buildings and vehicles with respect to self-generated solar energy and the usage of this energy depending on the size of the PV system and the stationary energy storage.

Figures 6 and 7 show the results of the simulation for the coverage of the household and EV power consumption by solar energy for two exemplary use cases, one commuter (25 km one way) and one household vehicle. The consumption profile of the building is based on a single-family house with a conventional heating system for a family of four.

The larger the PV installation and the storage battery, the higher the proportion of the energy demand that can be covered by solar energy. This proportion can be increased using an energy management system.

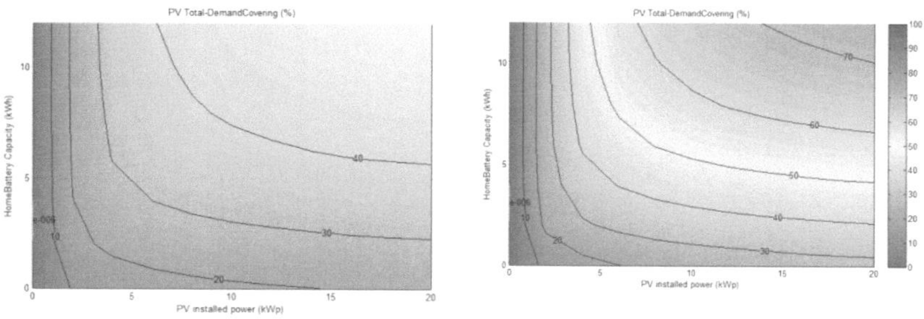

Figure 6: Commuter (left) and household vehicle (right) without energy management

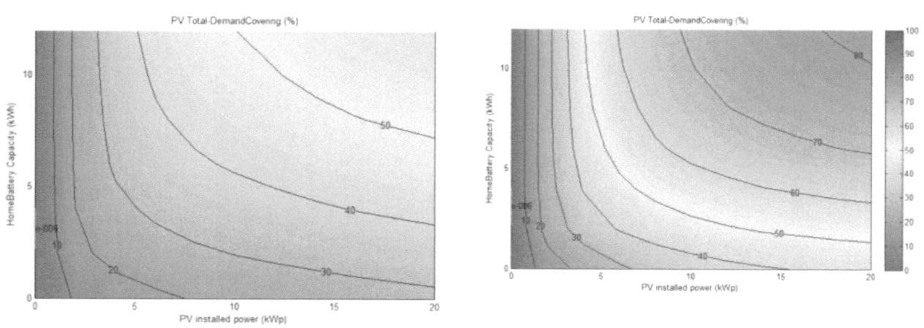

Figure 7: Commuter (left) and household vehicle (right) with energy management

It also becomes clear that the usage of the EV has a large impact on the solar coverage. The case of a household vehicle for example, which is used with an 8 kW PV system and a 6 kWh storage battery, demonstrates that the solar coverage can be increased from 50 % (immediate charging) to 60 % when an energy management system is used for intelligent charge planning.

Figure 8 shows the cost potential of integrating EVs into an energy management system. A conventional vehicle costs around 820 € yearly, but around 27.5 % can be saved by using electricity instead of more expensive fuel. A solar installation brings further energy cost reductions of 27 %. Intelligent charge management results in a larger coverage of the energy demand by solar power, which can lead to cost savings of about 150 € yearly. This corresponds to a cost reduction of 35 % compared to immediate charging with an existing PV system.

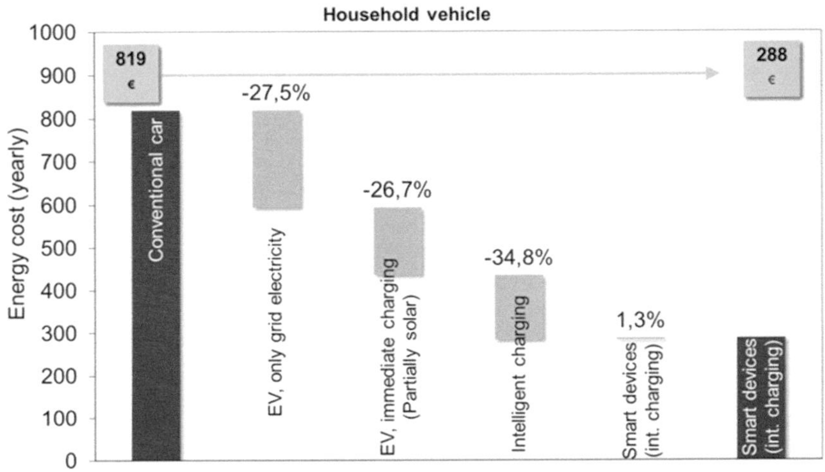

Figure 8: Cost saving potential for charging of electric vehicles

The energy that can be used for electric mobility and the resulting self-covered mobility potential are influenced by the building parameters (PV power, storage battery, heating system), but more importantly by the correlation between the user profile, i.e. the plug-in times and the energy demand, and the daily and seasonally varying energy supply. This impact is significant even if a stationary storage battery is used.

Figure 9 shows the theoretical mobility potential from off-grid energy sources at the best day of the month as well as the practical maximum solar mobility each month considering the plug-in times and charging demands. The monthly average is even lower, because of e.g. weather and usage variation. Nonetheless, on average 46 % of the yearly energy demand for mobility can be covered by solar power in the simulated case.

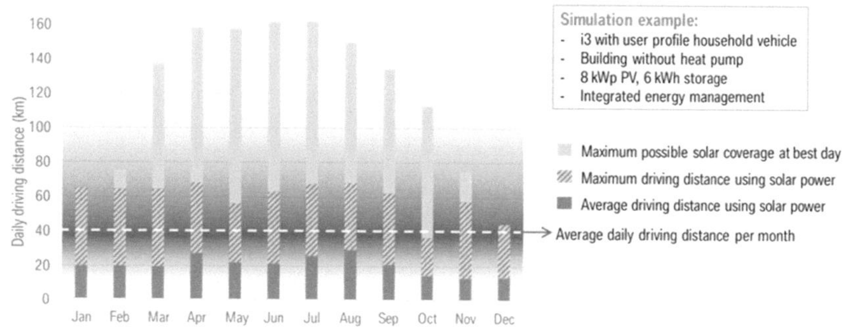

Figure 9: Solar mobility potential

4.3 Results from piloting in real operation

Within the scope of the project e-Mobilie, the concept described in chapter 4.1 and shown in figure 6 was first developed and tested using a test bench and then prototypically demonstrated in a single-family house from January 2015. A 7.8 kW PV system and a 10 kWh battery storage are installed on the building. The energy management system was implemented on a modified Sunny Home Manager, developed by the project partner SMA. A BMW ActiveE was used as the electric vehicle. It communicated via a wall-box as gateway directly with the energy management system. The vehicle was mainly used for commuting to work.

Figure 10 summarizes some of the results: Within a year the own use of PV power varies between 51 % in July and 99 % in December. This means that on average around 64 % of the generated electricity was used domestically. The solar demand coverage is reversed and averages to 46 %, with a maximum of 87 % in July and only 14 % in December. Naturally, the EV can use more solar electricity in summer. These observations support the simulation results.

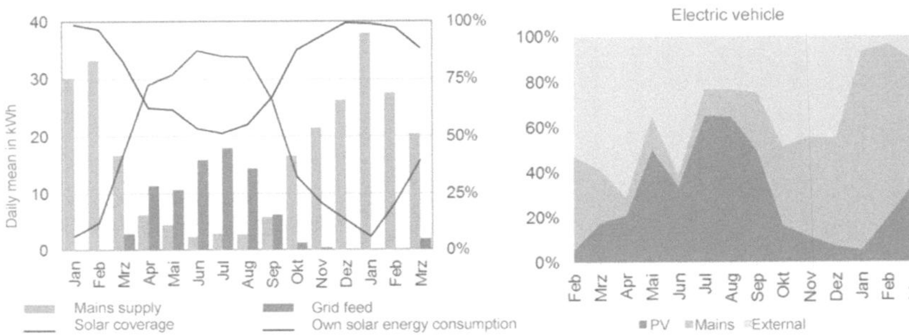

Figure 10: Energy generation and solar coverage of the entire household including EV (left) and coverage fractions of the EV (right)

5 Intelligent charging solutions as enabler for EV market success

5.1 All-round services and product solutions for electric vehicle customers

The current market penetration of electric vehicles is still far away from the targets set by the government. In order to reduce purchase barriers for private customers and fleet operators it is necessary to complement convincing EVs with all-round services and product solutions to simplify the transition to electric mobility:

- Consulting services for charging infrastructures and installation
- Electricity offers based on sustainable generation
- Charging concepts that intelligently interact with the local energy system
- Integration of electric mobility with modern digital services

Individual elements of the BMW ‘360° Electric’ services meet these customer needs. Digital charging services, which address different use cases and customer requirements, will soon be available for all private users and fleet operators using E-mobility.

Figure 11: Product range for digital charging services

5.2 Standardisation

The further development of standards concerning charging processes and the communication with control units are a major prerequisite for offering the services described in 3 and 4 in a broad and future-prove way.

The ISO15118 standard forms the basis for implementing system interactions as shown in figure 5. Concept development and piloting in projects such as the E-mobility showcase projects significantly contributed to developing and validating the ISO15118-based communication channel between vehicles and charging points.

Together with the EEBus initiative, use cases and data models were developed to evaluate the integration of charge planning processes into EEBus-based home energy management and home automation systems. These models will be used to implement a connection of vehicles to these systems in conformity with the ISO15118 standards.

5.3 Regulatory framework

The German government sees digitalization as an important aspect for a protection of the network stability during the energy transition, which should be regulated. The 'law on digitalization of the energy transition' ('Gesetz zur Digitalisierung der Energiewende') was designed for this purpose, which is currently being validated by the

Federal council and the German parliament. The law requires an intelligent measuring system, which is classified as a critical part of the network infrastructure. This smart meter gateway (SMG) forms the only interface between the network operators responsible for stabilizing the network and the controllable consumers.

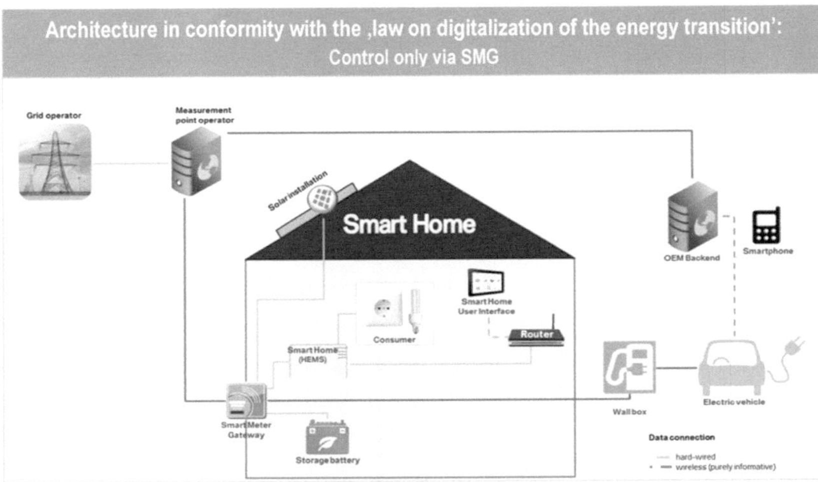

Figure 12: System architecture with smart meter gateway in conformity with the 'law on digitalization of the energy transition'

The compulsory SMG will be introduced progressively, amongst others depending on the household consumption. As users of electric vehicles have a larger energy consumption, they might be affected by the conversion early on from 2017.

The effects of the given system designs, external interventions via the SMG and possible functional interactions with local energy management systems such as HEMS or fleet charging systems are currently not fully ratable. The German Association of the Automotive Industry (VDA) therefore demands an extension of the phrasing, which assures a non-discriminatory control of charging processes and the protection of energy management solutions that are existing or will be implemented before 2020. In this way the intended customer benefits can be implemented through short-term product offers and thus help reduce the load on the power grid.

6 Summary

Intelligent charging control systems that interact with energy systems of the vehicle operator showed in simulation and real life testing that they can significantly contribute to satisfying the needs of both private customers and fleet operators:

- Minimize electricity cost through usage of self-generation and minimize capacity charge components of energy costs
- Securely comply with load limits and regulative connection conditions
- Maximize renewable energies in the consumed energy mix
- Avoid peaks in grid supply and grid feed-in

The BMW group develops attractive energy management product offers to make these advantages accessible for their customers.

PV energy and electric mobility – driving forces of the energy transition

Detlef Beister / Dr. Torsten Leifert

SMA Solar Technology AG
Sonnenallee 1, 34266 Niestetal, Germany
+49 561 9522 424276
Detlef.Beister@SMA.de
www.SMA.de

Developments in eMobility and Photovoltaics are Currently Lagging Behind Expectations

eMobility is the global key to an environmentally friendly transformation of mobility and, in Germany, it is part of the energy transition. The use of electric vehicles, particularly in combination with electricity from renewable energies, generates significantly less carbon dioxide. Thanks to their energy storage systems, electric vehicles will also be able to compensate for fluctuations in wind and solar power in a smart grid and thus support the development and market integration of these volatile energy sources. Germany, however, is trailing behind its stated political objectives in the field of eMobility (Fig. 1), according to which the number of existing vehicles should have reached 100,000 by the end of 2014. The actual figure was 24,000 [1].

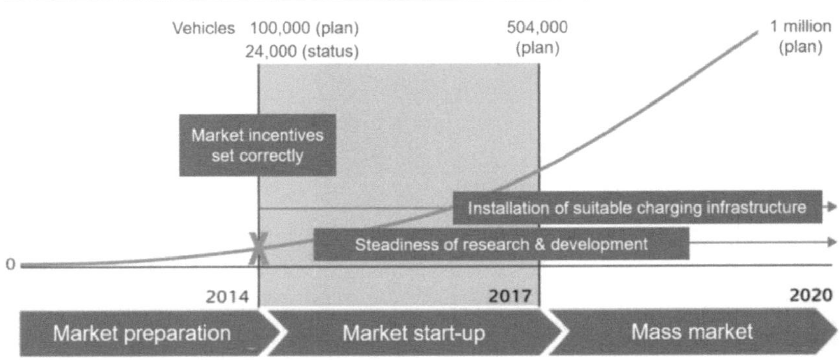

Fig. 1: Market development target curve 2010–2020 (main points for market start-up) [1]

Solar and wind energy are the most important renewable energy carriers for the energy transition. In 2015, they already accounted for 19.3% of gross power generation in Germany [2]. However, at 1,400 megawatts in 2015, for the second year running, photovoltaics (PV) were over 900 megawatts below the expansion corridor of 2,400 to 2,600 megawatts specified by statutory regulations [3] Fig. 2).

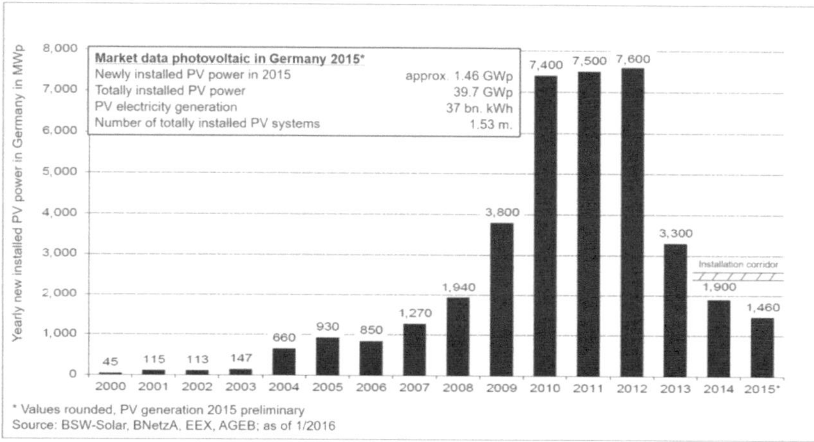

Fig. 2: Market data for photovoltaics in Germany [4]

In Germany, both eMobility and photovoltaic development are currently lagging behind expectations.

In Both Cases, Development Can be Boosted by Using Solar Energy for Charging as Part of an Intelligent Energy Management System

As eMobility only makes sense if the energy used comes from renewable sources, eMobility and photovoltaics are closely linked. In fact, in the private sector, there are many electric vehicle owners interested in intelligent solar charging as it ensures long-term cost stability for a significant portion of the mobility energy demand and makes photovoltaics even more attractive due to their higher self-consumption quota [5].

Renewable power generation is subject to variations in nature. Further developing the availability of highly flexible loads and storage systems therefore plays a key role for this form of power generation. Electric vehicles are characterized by their ability to receive and release high power levels, their heightened responsiveness and their in average rather long standstill periods – which makes them particularly fitting for short-term grid optimization [6].

The scenarios described presuppose integration of the charging infrastructure into the local energy management system of private and semi-public buildings, where the majority of the charging points will be located long term (Fig. 3) [1].

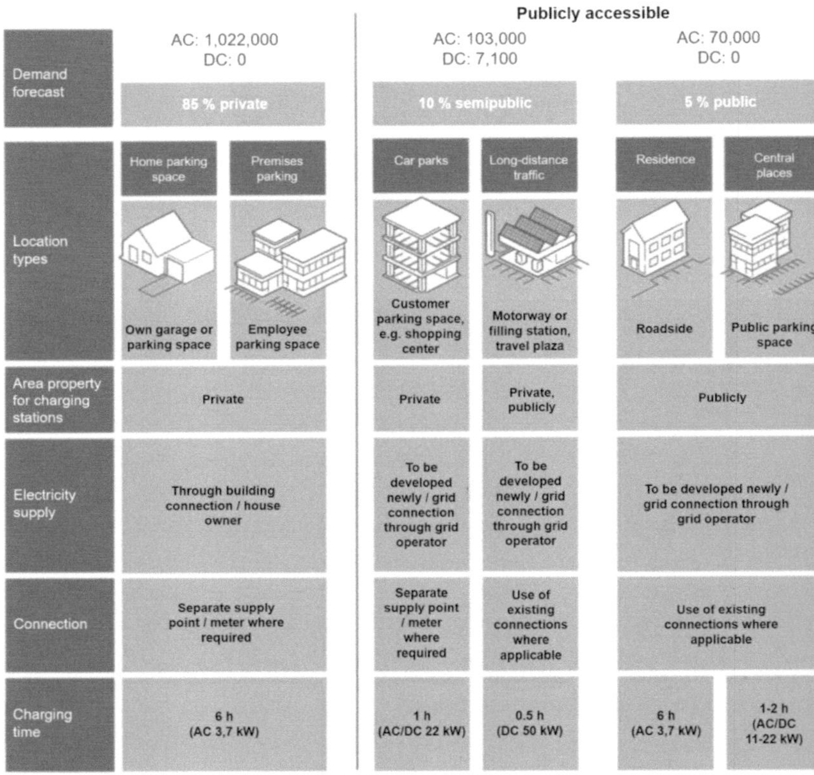

Fig. 3: Charging strategy 2020 based on the Pro scenario [1]

In turn, these buildings will eventually have to be integrated into higher-level structures (such as virtual power plants) to involve their flexibilities resulting from the combination of generators, loads and storage systems used as grid support in a smart grid, and use excess and required energy for trading on the smart market [7].

To date motivated by the aim of using as much self-generated photovoltaic power as possible, the focus has been on home installations. Now, however, companies with different requirements are also coming to the forefront if, for example, clever integration of photovoltaics and load control results in a reduction in grid peak loads and the associated demand charge. This makes it possible to save on energy costs while improving the eco-balance.

The Combination of Photovoltaics, eMobility and Energy Management Opens up New Business Models

In the following research and pilot projects, the individual aspects of such combinations were and are being analysed in detail:

1. "INEES" beacon research project
   - Integration of mobile batteries into the energy market
   - Intelligent grid integration of electric vehicles to provide positive and negative secondary balancing power
   - Field experiment with 20 vehicles and 25 bidirectional charging stations
   - Analysis of technologies, business processes, benefits to the energy industry and customer acceptance
   - Conducted by Fraunhofer IWES, LichtBlick, SMA Solar Technology, Volkswagen
   - Supported by the Federal Ministry for Environment, Nature Conservation and Nuclear Safety
2. "e-MOBILie" research project – eMobility showcase
   - Energy management for charging with as much PV energy as possible
   - eMobility in a single-family home and an intelligent garage
   - Researching potential system architectures and the potential for the connection of photovoltaics and eMobility
   - Conducted by BMW Group, IfE Institute for Energy Economy and Application Technology, SMA Solar Technology, TUM Center for Sustainable Building
   - Supported by the Federal Ministry for Environment, Nature Conservation and Nuclear Safety
3. "e-Optimo" pilot test
   - Optimized home energy usage through intelligent energy management
   - Charging electric vehicles subject to variable electricity tariffs and in the context of decentralized power generation, electric and thermal loads and storage systems
   - Automatic selection of the most profitable energy consumption variant
   - Conducted by Allgäuer Überlandwerk, SMA Solar Technology, Volkswagen

The results of these projects present clear recommendations of action for the technology required and the general economic conditions needed. Below, a distinction is drawn between private and commercial solutions due to the fundamental differences exhibited in multiple areas.

System for private use

In a nutshell, a complete system consisting of eMobility, photovoltaics and energy management currently available on the market could look as follows (Fig. 4):

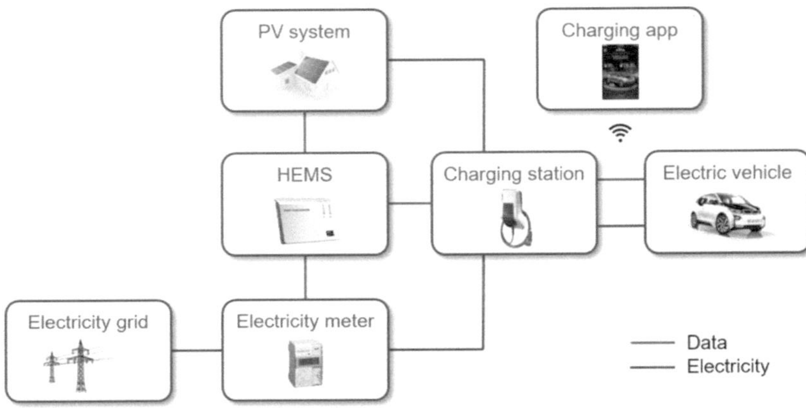

Fig. 4: Example of a system for private use

The standard PV system is equipped with a communication-enabled inverter. The energy meter monitors purchased electricity and/or grid feed-in and also has an appropriate, local communication interface. At one end, the charging station with controllable charging power communicates the maximum possible charging power to the connected electric vehicle and conversely receives the charging power from the electric vehicle needed at that time. It measures the active flow of current via an integrated energy meter and, at the other end, delivers all relevant data about the charging process to the Home Energy Management System (HEMS).

The smart phone charging app communicates with either the electric vehicle or the charging station, depending on the specific system version. Among other things, the app can be used to enter limited – yet important – information about the charging process, such as the amount of energy required before the next departure, the departure time and whether the electric vehicle should be charged using only accumulated PV excess. In addition, it visualizes the charging process for the user.

As the final step, the HEMS processes the data from the PV inverter and the energy meter. It monitors the relevant energy flows and provides a recommendation for the charging time period and the appropriate charging power to the charging station based on the charging information provided by the user. It takes into account the entire home consumption situation as well as PV generation and, where applicable, the current variable electricity tariff. This charging plan is then integrated into the process of further optimizing the energy efficiencies of the building, including the charging plan for the stationary thermal and electric storage system. Ultimately, the battery management system in the electric vehicle decides on the amount of charging power that is actually needed.

In addition to the small amount of basic conditions entered by the user, the electric vehicle is charged conveniently and automatically. This is also important because the efficiency of the system greatly depends on user acceptance. The more the system operates in the background (though as intended), keeping intervention in daily life to a minimum, the greater the user acceptance is. Of course, the user cannot be spared from plugging in the vehicle whenever possible. With vehicle-to-home and vehicle-to-grid applications in particular (see below), it goes without saying that vehicles need to be plugged in, even if actual charging is not required at the moment. Once users recognize the benefit of this system, any initial scepticism about the amount of new technology quickly disappears and this unusual approach becomes habit. In the e-MOBILie single-family demo home, the daily average for the amount of time the vehicle was plugged in was 13 hours.

Optimization targets for the private use system

If the vehicle does not need to be charged immediately to full capacity, the main HEMS optimization targets for this solution are to achieve the highest possible PV self-consumption quota and to proficiently consider a variable electricity tariff. In doing so, flexibility can result from both the charging time period and the charging power. Furthermore, under certain circumstances, dynamically limiting the maximum charging power will prevent overloads at the grid-connection point as the maximum possible power of a charging station can represent a significant portion of the maximum power supply from the grid technically possible for a house connection. The aim is to reduce energy costs while improving the life cycle assessment. And all this should be achieved not just without inconveniencing the user but, even better, by increasing convenience.

In the future, in addition to the optimized charging that has been described so far, we will also see the gradual use of vehicle batteries in smart homes – also known as vehicle-to-home, or V2H for short – and on the smart grid or the smart market – also known as vehicle-to-grid, or V2G for short (see below).

Technical specifications for private use

These optimization targets can be achieved with varying levels of outlay. Typical specifications could look as follows:

Table 1: Different system specifications for private use

		Threshold control	Algorithms, predictions	Simple communication between electric vehicle and charging station (e.g., ISO 61851)	Enhanced communication between electric vehicle and charging station (e.g., ISO 15118)	Proprietary communication between HEMS and charging station (e.g., SEMP)	Standardized communication between HEMS and charging station (e.g., EEBus)	Sensor technology to monitor the grid-connection point	Bidirectional electric vehicle	Bidirectional charging station
1.	Basic solution	✓	×	✓	×	✓	×	×	×	×
2.	Grid connection monitoring	✓	×	✓	×	✓	×	✓	×	×
3.	Simple energy management	✓	×	✓	×	✓	×	×	×	×
4.	Enhanced energy management	✓	✓	✓	×	✓	×	×	×	×
5.	Vehicle-to-home unidirectional	✓	✓	×	✓	×	✓	✓	×	×
6.	Vehicle-to-grid unidirectional	✓	✓	×	✓	×	✓	✓	×	×
7.	Vehicle-to-home bidirectional	✓	✓	×	✓	×	✓	✓	✓	✓
8.	Vehicle-to-grid bidirectional	✓	✓	×	✓	×	✓	✓	✓	✓

The farther down the table a specification is, the greater is the level of outlay involved but also the achievable benefit. Another important aspect in this regard is the future use of distributed energy management instead of a hierarchical approach. This means that the HEMS works together with local intelligent systems (e.g., in the electric vehicle or the home heating system, to achieve an even better overall result).

A targeted result of integrated energy management may look as follows (Fig. 5):

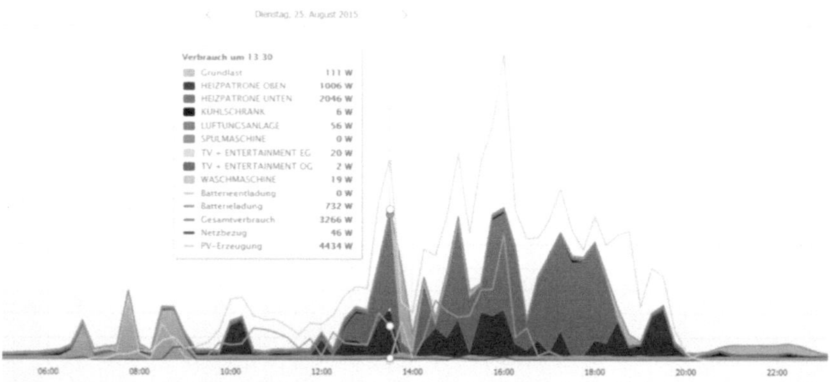

Fig. 5: Energy balance of individual devices in a building with HEMS

Here, you can see how, on a cloudy day, the HEMS distributes the connected devices under the PV generation curve (yellow line) so that purchased electricity (red line) is largely avoided.

Current status and further development for private use

For both private and business use, there are currently, more or less, only proprietary solutions from levels one to four on the market. To date, level four has produced the best results with its forward-looking and automatic optimization of the charging process that takes into account user needs, vehicle requirements and its generation and consumption predictions. Naturally, the mass market can only be addressed with a high degree of international standardization. In the E-Mobility Working Group in the EEBus initiative, companies from the PV industry are working on this standardization with the German Association of the Automotive Industry (VDA) and charging station manufacturers.

We are currently on the brink of level five, which uses standardized interfaces as a significant step forward. e-MOBILie has already developed the concept and demonstrated the proof of concept for distributed energy management based on ISO 15118 and the Simple Energy Management Protocol (SEMP). Levels six to eight will gradually become a reality over the next five years or so. As a result, the emergence of these mass-market solutions fits perfectly with the development of a tailored charging infrastructure in the market start-up phase (Fig. 1).

Although more complex solutions only tend to become widely available on the market later on, it is likely that, in the long run, all of the specifications will be on the market at the same time to ensure that various needs and interests can be met. Users must decide on their preferred outlay/benefits ratio and therefore the nature of their ideal system for themselves.

Costs and benefits of a system for private use

Initial investments are required for the purchase of an HEMS and other related components. Furthermore, the additional energy consumption caused by the complete energy management system also involves operating costs. Over the service life of the system, both of these need to be paid off. The e-MOBILie single-family demo home has demonstrated that, in the medium term, the pay-back period for purchasing and operating an integrated energy management system can be under two years for modern, single-family homes with controllable loads.

In the e-Optimo pilot, it was possible to use an HEMS to increase the PV self-consumption quota and charge the battery at cheaper tariff times. The average range of savings that can be achieved this way are as follows:

- The average yearly saving for a family household (4,000 kWh) ranges from €60 to €90 with a 4 kWp PV system and €80 to €130 with a 7 kWp PV system (south facing, reference year 2011)
- The average yearly saving for a single person household (4,000 kWh) ranges from €65 to €100 with a 4 kWp PV system and €85 to €140 with a 7 kWp PV system (south facing, reference year 2011)
- A differential tariff with an energy price between 24.39 and 29.62 euro cents per kilowatt hour, depending on the time of day and the season, and a basic charge of €10 per month bring further savings of €10–€15 per year for eMobility customers with a mileage of 10,000 km per year.

However, the actual PV self-consumption quotas and the potential for savings largely depend on the building features, the size and properties of the PV system, the size of the existing electric and thermal storage system and the individual consumption habits of the household. Notably, the impact of the electric vehicle is largely determined by whether it is at home a lot during the day (e.g., "family car") or not (e.g., "commuter vehicle"). Frequent drivers need to use the vehicle very often and therefore must be able to charge their vehicle at a high charging power level to enable short time shifts, otherwise the vehicle has to be charged immediately to ensure availability. The opposite is the case for infrequent drivers. As the electric vehicle is not driven very often, although there is a large amount of time available for solar charging, the amounts of energy that can be collected will only be very low.

It is often said that the impact of electric household appliances, such as washing machines, driers and dishwashers, on savings is rather low. Depending on the driving profile and daily mileage, electric vehicles offer a much higher potential for savings. The highest potential, however, is offered by space heating and domestic hot water heating using heat pumps or electric direct heating.

In addition to the objective, purely financial benefits of using an HEMS, there are also several subjective benefits that are difficult to quantify, which are attributable to the high self-sufficiency quota and the overall improved eco-balance.

Market potential for private use

There is huge market potential for these types of solutions. Many single-family homes and their corresponding parking spaces are already suitable or can be made so with reasonable outlay. Furthermore, all of the equipment considered here and all equipment required for an energy management system can be installed not only in new constructions but also in existing buildings. Even in a new construction, it is usual to continue using existing household appliances, for example. In this context, the biggest difference between new constructions and existing buildings may be the outlay required for interlinking the devices. While today there is typically network cabling installed in all rooms of new constructions, in existing buildings, solutions that use wireless communication (WLAN) or power-line communication (PLC) via the existing power lines are often resorted to. Usually, however, the availability of a corresponding network can be assumed for other reasons (e.g., ICT, convenience, security, multimedia), meaning that no additional costs are incurred for the energy management system.

Unfortunately, this type of system will be difficult to transfer to multiple-family dwellings in the foreseeable future for various reasons.

System for business use

All of the considerations for the private solution regarding the various aspects of technical implementation, outlay and benefits, existing solutions and the roadmap for the next few years also apply, to a large extent, to the business solution.

It goes without saying that the exact technical specifications of a complete business system including eMobility, photovoltaics and energy management vary significantly depending on the application. In a nutshell, a system currently available on the market for a company parking garage could look as follows (Fig. 6):

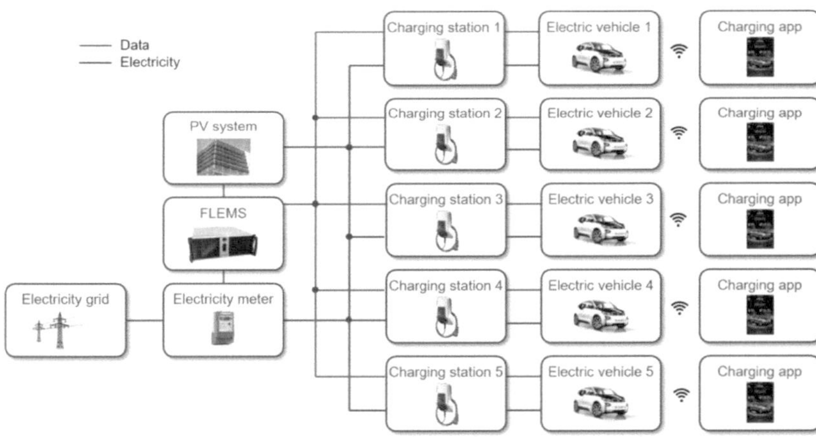

Fig. 6: Example of a system for business use

The components used are fundamentally very similar to the components used for the private solution. The PV system is generally bigger, yet still commercially available, and is equipped with one or several communication-enabled inverters. The energy meter looks somewhat different than the one used in the private sphere, but it also monitors purchased electricity and/or grid feed-in and also has a suitable, local communication interface. At one end, the charging stations with controllable charging power communicate the maximum possible charging power to the connected electric vehicles and conversely receive the currently required charging power from the electric vehicles. They measure the active flow of current via an integrated energy meter and, at the other end, deliver all of the relevant data about the charging processes to the Fleet Energy Management System (FLEMS). Again, the smart phone charging app acts as the front end for the user. At this point – for larger installations, in particular due to reasons of cost and complexity – it may be useful to replace the individual communication function of each charging station with the connected car and/or the FLEMS with a central, back-end communication system.

In this environment, the FLEMS occupies the role of the HEMS and processes the data from the PV inverters and the energy meter. It monitors the relevant energy flows and provides a recommendation for the charging time periods and the appropriate charging powers to the charging stations based on the charging information provided by the users. In doing so, in the same way as the HEMS, it also takes into account PV generation, the current variable electricity tariff amount (where applicable) and the entire consumption situation in the building. Its most important optimization target, however, is to ensure that the maximum amount of purchased power in the building remains as low as possible (peak shaving) because the purchased electricity price is calculated according to this. These charging plans are then integrated into the process of further optimizing the energy efficiencies of the building, including the charging plan for the stationary thermal and electric storage system. The priority here is reducing energy costs. Additionally, in many cases the FLEMS can be linked with an existing fleet management system.

In addition to the small amount of basic conditions entered by users, the electric vehicles can be optimally charged conveniently and automatically as, in this case too, the efficiency of the system greatly depends on user acceptance. Most of the 22 participants in the e-MOBILie parking garage demo believe that there is no reason not to use a charging management system. They recommend that eMobility be only offered as part of a package with the charging management system and an allocated charging area. There is, however, also the view that the package must offer the user personal benefits if they have to enter the departure times. For some, this simply means a green conscience. Others would like to receive a direct and positive cost balance such as cheaper charging than at home or cheaper charging than without scheduled charging. At 77%, most of the participants indicated that they always charged the vehicle with an active departure time set. In addition, 48% indicated that they never left before the assigned departure time. Compared to immediate charging, therefore, the grid peak loads were reduced by 7.8 kW and self-sufficiency increased by 20%.

Vehicle-to-grid integration

Significant findings and recommendations for integration of vehicles into the secondary balancing power market have been acquired from the INEES project, according to which the potential for revenue and efficiency greatly depends on the regulatory framework conditions. Limited offer periods and daily tender periods would increase the potential. Equalizing the fees associated with pumped storage power plants would significantly increase the potential for revenue.

Important prequalification requirements have already been met. The fact that vehicle users and the energy industry use the same communication network for the vehicle still is a major obstacle.

The required power measuring accuracy is achieved from 200 vehicles. Prediction and determination of the safe operating reserve that can be supplied is achieved from 500 vehicles. The valid minimum offering of 5 MW is achieved from 5,000 vehicles.

In principle, electric vehicles are well suited for use on the secondary balancing power market thanks to their fast and precise activation. Nevertheless, it has been demonstrated that the stress on the vehicle battery in connection with secondary balancing power is linked to a significant loss in value, which does not currently appear economical for this type of model. Preliminary studies of other areas have revealed that the situation may be considerably better in relation to primary balancing power, but this is yet to be explored.

Irrespective of whether V2G or V2H are concerned, user settings and behaviour significantly affect the storage potential that can be used in this way (Fig. 7).

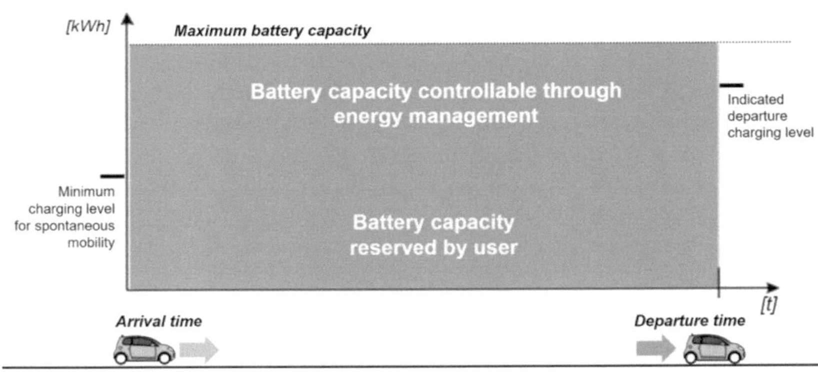

Fig. 7: Changes in battery capacity usage over time

If the battery charge level is below the minimum charge level for spontaneous mobility when the vehicle arrives, the battery is recharged to this level immediately. Depending on the departure time and charge level entered, the battery is then re-charged to the specified level in time. Both of these charging processes are represented by the red area in the figure and are not available for flexibility options.

In the figure, the delta between the red area and the maximum battery capacity is represented by the green area. This is available for flexibility options and can be quite extensive depending on the specific driving behaviour and the defined settings. The INEES project observed that a value set by users contributed, on average, about 30% to energy efficient usage. It is likely that this value will continue to increase over the years with the growing capacity of vehicle batteries.

Current technical and regulatory challenges

In addition to the aforementioned expansion of communication protocols related to energy management, there are other gaps in and changes that need to be made to the current regulations and industry standards. Examples include replacing the existing requirement for galvanic isolation between the charging station and the electric vehicle with other, more cost-effective solutions, as is already common practice for photovoltaics; significantly reducing the connection time currently required by the technical specifications of the distribution grid operator from 120 seconds to 1 second at the start of feed in; and replacing the recorded load profile measurement with a more cost-effective process such as smart meters or device measurement.

There are therefore both technical and regulatory challenges to overcome. In addition to the aforementioned points regarding the use of vehicle batteries on the operating reserve market, in relation to this, the "digitization of the energy transition draft law" currently being discussed by the German Federal Ministry for Economic Affairs and Energy is particularly noteworthy. This package envisages the introduction of a "smart measurement system" for buildings with PV systems from 7 kWp as of 2017. An assessment is also being conducted to find out whether installation requirements for "smart measurement systems" should be provided for self-consumption configurations in general ("prosumers"). This would affect the majority of existing PV systems and all new PV systems. Efforts are being made to transform this "smart measurement system" into the single, compulsory communication interface to a decentralized energy system with a smart meter gateway administrator as a central entity, independent of the current grid status. All generators and energy-relevant loads, including the electric vehicle, would be connected via a "control box" to allow the grid operator to control generation and/or consumption if required. An enhanced energy management system, as well as the vehicle-to-home and vehicle-to-grid applications described in this article, would then no longer be implemented or, at least, would no longer be economical to implement.

For the benefit of everybody involved, an open ICT infrastructure needs to be standardized for smart energy management (e.g., use cases, data structure, smart meter gateway with suitable customer interface, CEMS gateway with suitable protection profile). The logical completion of the energy transition in an energy internet [8] requires the distribution of intelligence as opposed to central control. An appropriate model is a cell structure with theoretically any number of different levels. Every cell can act autonomously to a certain extent, but they are also always involved in the processes of the upper levels [9]. Current technological trends, such as the Internet of Things and the Smart Home, support this approach. One example of use of this type of distributed intelligence is the automatic negotiation of the aforementioned schedules between the higher-level and the local energy management system. Only a system with distributed intelligence instead of a centralized structure will be robust and manageable in the long term. Therefore, a corresponding strategy must be pursued.

In an ideal case, a smart grid and smart market scenario would not be regulated and compulsory for all prosumers. Metering point operators would be neutral and would not hamper the competition between providers of energy services. Instead, participation would be voluntary and incentive-based. In this case, to ensure customer acceptance, the schedules must be designed in such a way that considers basic individual situations, ensures that customer convenience is always guaranteed and does not interfere in the customer's daily routine but in fact, in an ideal case, supports it. Energy data and flexibility would remain in the property of the end user.

Another obstacle that has to be overcome is modification of the regulatory framework to expand the scope of using variable electricity tariffs as an incentive for private electricity consumers. This has not been successful up to now mainly because, apart from the reduced concession fee, utility companies cannot generate cost advantages due to a load transfer for scheduled customers in accordance with the strict H0 standard load profile (SLP). The measurement, invoicing and settlement according to meter readings recorded in 2013 as per Section 12, Paragraph 4 of the German Utility Grid Access Regulation (Stromnetzzugangsverordnung – StromNZV) could suggest the possibility for allowing higher monetary incentives for load transfer than the SLP procedure. This could involve the electric utility companies purchasing the predicted load for customers based on individual load profiles to reduce their demand for expensive energy from the operating reserve and ultimately make it possible to offer household customers attractive, variable tariffs. However, the precise configuration of the rules for the required market communication has not yet been specified and accounting based on a series of meter readings determined every 15 minutes ("meter reading profile accounting") has still not been considered in MaBiS 2.0 (the "Market Rules for the Implementation of Electricity Balancing Group Billing" of the German Federal Network Agency) [10].

Energy market model

The energy market model is another important topic. For the European internal energy market to develop in the future, it is essential that the flexibility of the system be increased. In addition, the behaviour of users of grid state variables who are connected to the utility grid can be influenced to respond to various grid states in the most beneficial way possible for the grid. One of the best known approaches for distinguishing various grid states is the "traffic-light principle" of the German Association of Energy and Water Industries [11]. The basic aim of the principal is to regulate interaction between the smart market and the smart grid based on the grid states "green," "yellow" and "red". In this way, any interventions from the grid operator outside of the smart market are limited to ultima ratio decisions. To make participating in the energy market beneficial for everybody involved, new remuneration schemes for trading flexibility must be implemented, the grid payment system must be altered accordingly and a sustainable energy market model based on this traffic light principle must be introduced because, with the current regulatory framework, the CAPEX and OPEX for the new infrastructure needed cannot be offset. Furthermore, a clear aggregation structure needs to be defined and complemented by regional markets offering new electricity products, such as remuneration for supplying local flexibilities.

New business models

In all cases, several new business models are arising in this field, initially for the users of the electric vehicles mentioned above, aggregators, component manufacturers and ICT providers. At this point, the role of the installer, in particular, should be highlighted. To a large extent, these types of complex solutions are only successful if potential customers can obtain the solution from a single source. Nobody is better suited to this than a solar technology specialist, who can and should be the link between the solar industry, the automobile industry and users of electric vehicles.

Synergies of Photovoltaics and eMobility Will Accelerate Penetration of Both Technologies

The regulatory changes described above will make a number of appealing offers possible that would rapidly increase the available number of decentralized energy systems involving photovoltaics and eMobility as stable pillars of the energy transition for smart grids and smart markets and therefore ensure continued stabilization of the technical and economic system.

Pooling of mobile batteries in the gigawatt range can make a significant contribution to stabilizing the grid and generate revenue for everybody involved – in both the private and the commercial sphere. Integrating batteries into the local energy management system in smart buildings combined with photovoltaics and variable electricity tariffs improves the efficiency of the components concerned and has a positive effect on the current "price per kWh" obstacle for eMobility. It also increases the demand in the industries for both eMobility and photovoltaics, promotes the energy transition in Germany and sets milestones for global use of such concepts.

References

[1] National platform for eMobility: *"Fortschrittsbericht 2014" [Progress Report 2014]*, Berlin, December 2014

[2] BMWi website "Erneuerbare Energien auf einen Blick" [Renewable Energies at a Glance], retrieved on January 14, 2016

[3] German Federal Network Agency press release "Die Fördersätze für Photovoltaik-Anlagen werden erneut nicht abgesenkt" [Funding for Photovoltaic Systems Will Not Be Reduced Again], Berlin, December 30, 2015

[4] BSW: "Meldedaten PV Bundesnetzagentur 2014/2015" [German Federal Network Agency PV Reporting Data], December 2, 2015

[5] Navigant Research: "New EV Chargers Balance Solar, Home Power Demand", October 30, 2015

[6] Christian Höhle, Dr. Torsten Leifert: "Intelligente Nutzung von PV-Strom für die Elektromobilität als Wegbereiter für eine 100% erneuerbare Stromversorgung" [Smart Use of PV Energy for eMobility as a Forerunner for a 100% Renewable Electricity Supply], contribution to PV Symposium, Bad Staffelstein, March 2014

[7] Detlef Beister, Dr. Debora Coll-Mayor: "IKT-Lösungen zur Einbindung von PV-Anlagen in virtuelle Kraftwerke" [ICT Solutions for the Integration of PV Systems in Virtual Power Plants], contribution to ETG symposium "Von Smart Grids zu Smart Markets" [From Smart Grids to Smart Markets], Kassel, March 2015

[8] BMWi brochure "Entwicklung digitaler Technologien" [Developing Digital Technologies], Berlin, October 2014

[9] acatech study *"Future Energy Grid"*, Munich, February 2012

[10] PricewaterhouseCoopers blog article "Zählerstandsgangbilanzierung als Wegbereiter für variable Stromtarife?" [Meter Reading Profile Accounting as a Frontrunner for Variable Electricity Tariffs?], Frankfurt am Main, April 2014

[11] BDEW roadmap "Realistische Schritte zur Umsetzung von Smart Grids in Deutschland" [Realistic Steps for the Implementation of Smart Grids in Germany], Berlin, February 2013

Photovoltaics and electric mobility – potentials and reasonable integration into the power grid – data and facts gained from practical experience

Dipl. Ing. (Univ.) Hans Urban, Schletter GmbH

1 An interesting calculation

As an introduction, we want to take a closer look at the enormous potential of the electric mobility of the future that will be fuelled by renewable energy using the example of a specific solar plant that has been in operation since 2008. Even though storage and charging management surely still are a big challenge, the comparison of power generation and demand shows that renewable mobility basically is possible. In this context, it is also interesting to reconsider well-known reservations that electric cars hardly ever can be charged by solar plants at daytime in most cases, only have a limited range and are also too expensive. The debate should finally focus primarily on all the cases that already now can be solved with electric mobility and the implementation of these cases should be started. In view of the ongoing technological progress, there will be further potentials very soon.

2 Electric mobility as a tool for economic optimization by increased self-consumption

Currently, in most cases, PV plants can only be profitably realized if the self-consumption quota is high enough. This applies to craft businesses as well as to trade and industrial companies. The comparison of the different possible applications of solar power clearly shows that electric mobility with direct charging from the solar plant is definitely by far the economically most attractive use of solar power.

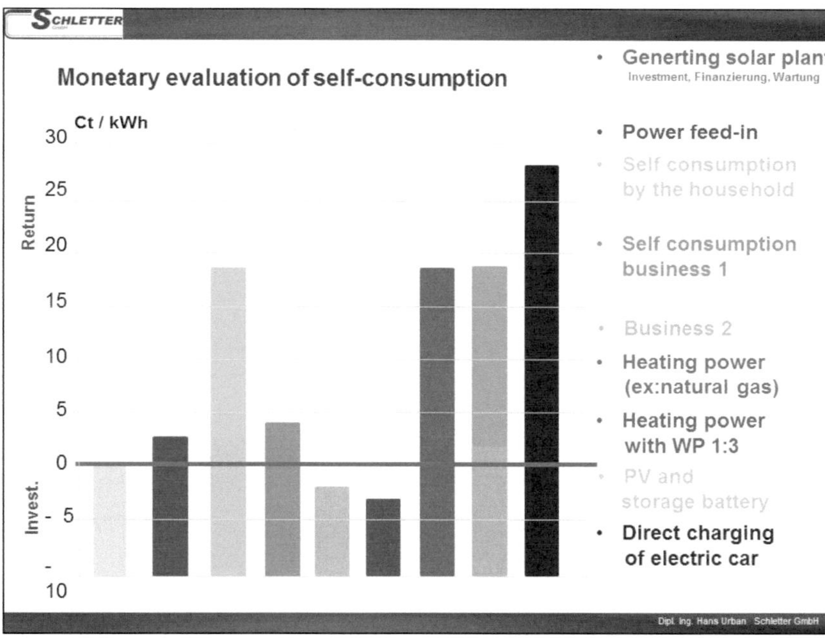

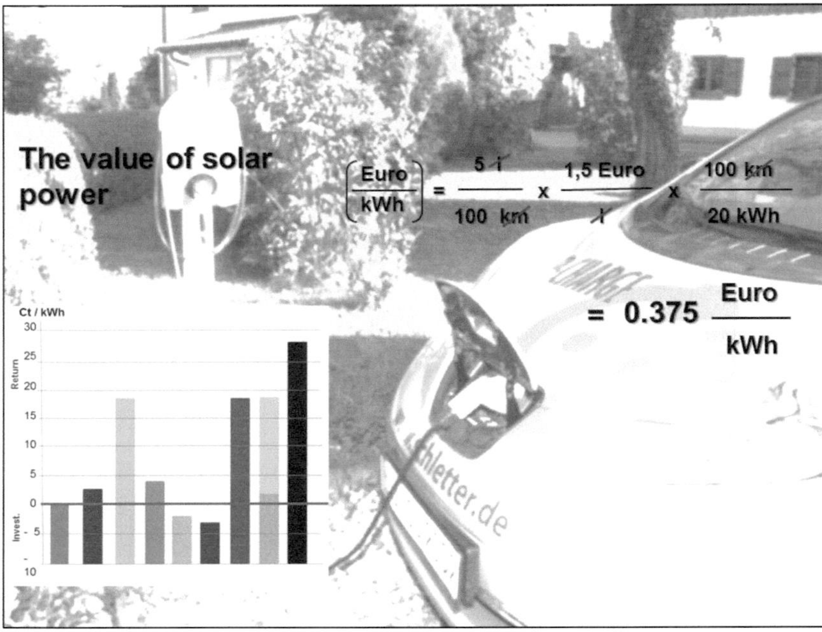
The value of solar power
(Euro / kWh) = 5 l / 100 km x 1,5 Euro / l x 100 km / 20 kWh
= 0.375 Euro / kWh
Ct / kWh
30
25
20
15
10
5
0
- 5
- 10
Return
Invest.

3 Sample 2012 – household without storage

Using the example of sample data from a household with a 10kW solar plant for self-consumption of solar power, the potential of electric mobility in combination with other appliances (power-consuming devices) in the household is shown. It is obvious that self-consumption quotas can be increased significantly when self-generated solar power is used for electric vehicles, as far as the electric vehicle can be charged completely or partially directly from the solar plant as often as possible. An optimized charging process that is able to adapt the charging power of the electric vehicle flexibly to the excess power that is available is the ideal mode of operation. This also applies if there is no solar storage system installed in the household, in ideal cases, an electric vehicle with controlled charging can even be used as a substitute for a solar storage system.

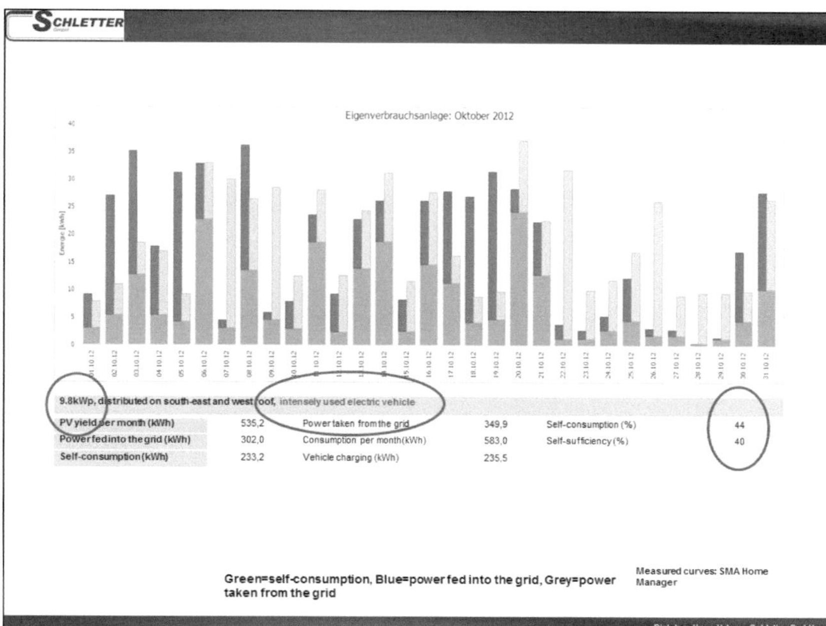

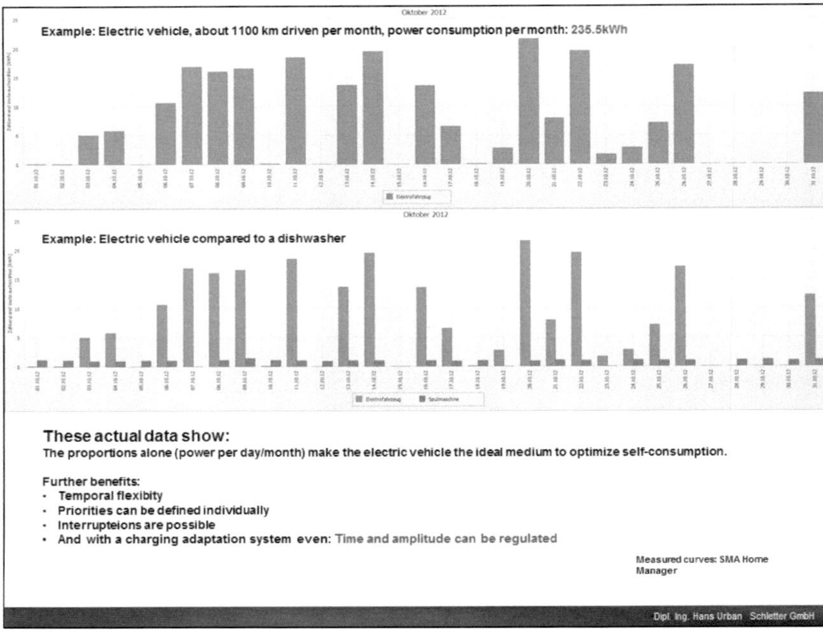
Oktober 2012
Example: Electric vehicle, about 1100 km driven per month, power consumption per month: 235.5kWh
Oktober 2012
Example: Electric vehicle compared to a dishwasher
These actual data show:
The proportions alone (power per day/month) make the electric vehicle the ideal medium to optimize self-consumption.
Further benefits:
• Temporal flexibity
• Priorities can be defined individually
• Interrupteions are possible
• And with a charging adaptation system even: Time and amplitude can be regulated
Measured curves: SMA Home Manager
Dipl. Ing. Hans Urban Schletter GmbH

4 Sample 2015 – household with storage 15 kWh

The combination of different systems in the household, especially storage system, electric vehicle and possibly also the use of solar power for thermal purposes is the most flexible "final stage", which provides many economical optimization options for the household and is also relevant regarding the national economy as costly expansions of the power grid are made redundant and combination of different systems also helps to reduce the capacities required for the provision of power. However, the debate should remain within a realistic range, because in our temperature zone, households will require power from the grid or from another source (for example from a block heat and power plant) for at least three months a year.

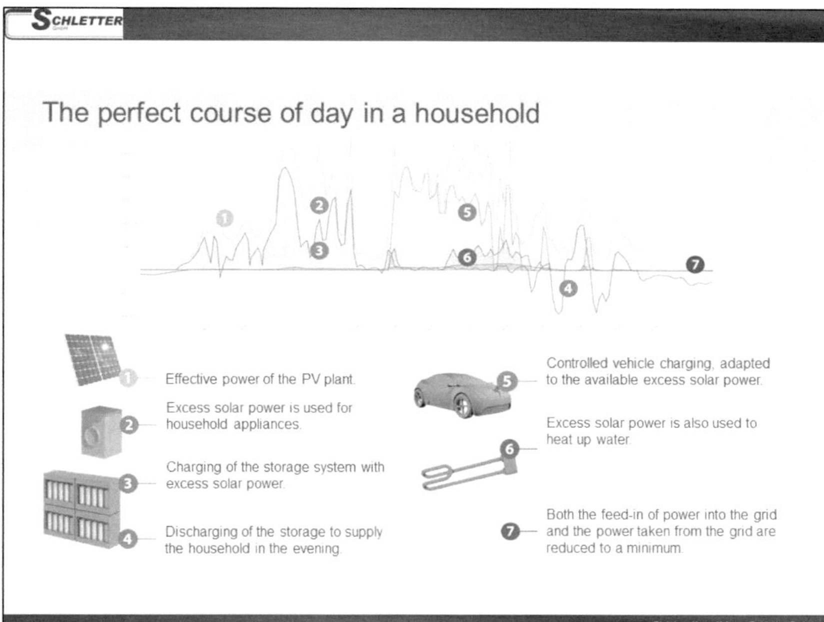

5 The household of the future – actual data

The realistic potential in the household of the future is shown both on the basis of a measured widely optimum daily routine and an evaluation of a period of 10 months.

Self-sufficiency quotas of 80% are also possible in the long run!

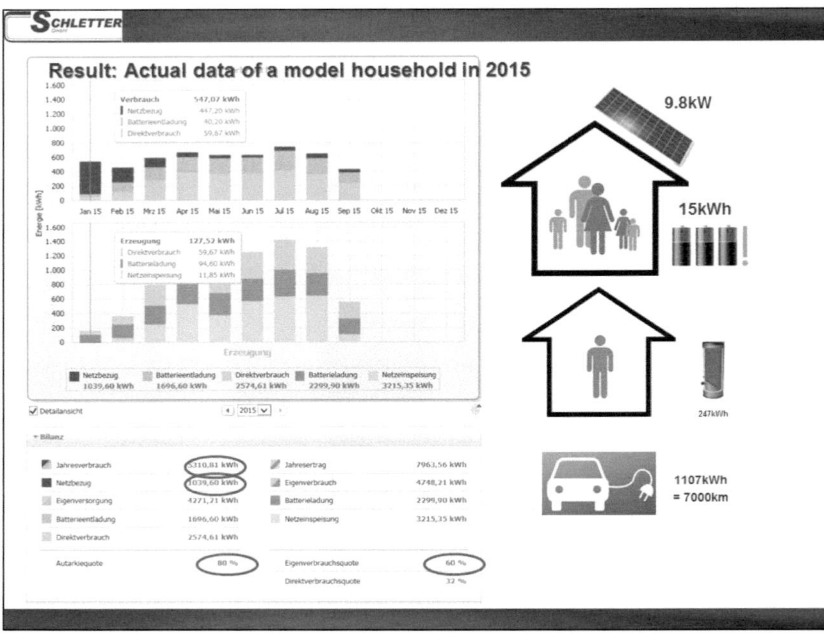

6 The business enterprise of the future – with integrated mobility

It goes without saying that electric vehicles are not only an option for private households, but also for trade and industrial companies. Actually, the potential is even bigger there, because electric mobility on the basis of solar power perfectly meets the requirements of service vehicles in many cases, and if you include the vehicles of the staff members in this scenario that are parked on the company parking lot for most of the day, the potential becomes almost limitless, and as power supply and power demand coincide most of the time, a far bigger self-consumption quota can be achieved by direct charging than with intermediate power storage.

7 Carports as an ideal completion of an overall concept

Where solar plants on roofs are not possible or already have been installed, the PV carport is an ideal completion. In this context, it is surely interesting that each solar carport parking space generates the power that an electric car needs to drive 10,000 kilometers. Apart from that, a company, a municipality or any other organization can make a visibly promote mobility on the basis of renewable energy.

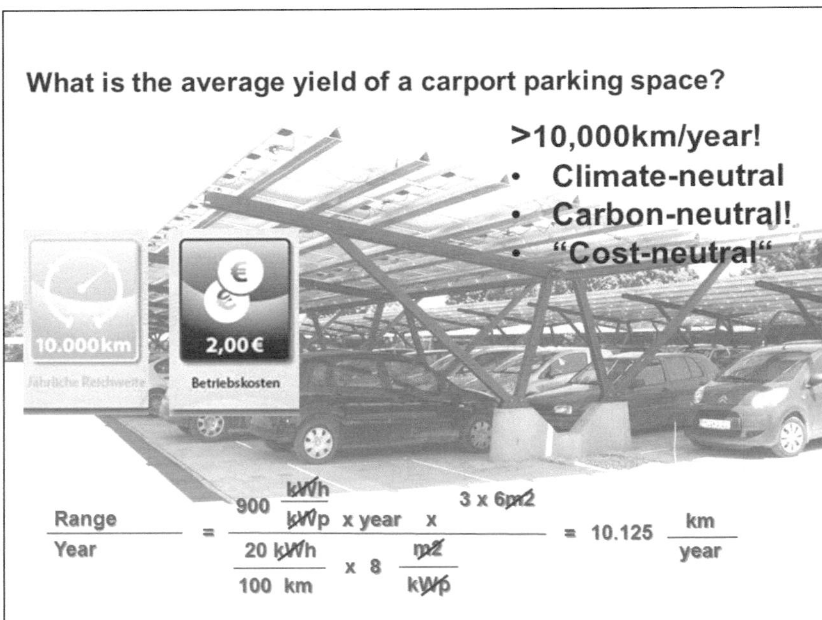

8 PV carports: Examples

Reference carport projects all over the world show how big the potential of this optimal combination of photovoltaics and electric mobility is and how big it still can become!

SCHLETTER
• 600 kWp - 2012
Dipl. Ing. Hans Urban Schletter GmbH

SCHLETTER

Photovoltaics and electric mobility

Potentials and reasonable integration into the grid

Data and facts gained in practice

Dipl. Ing. Hans Urban Schletter GmbH

SCHLETTER

1. Introduction:
An interesting calculation

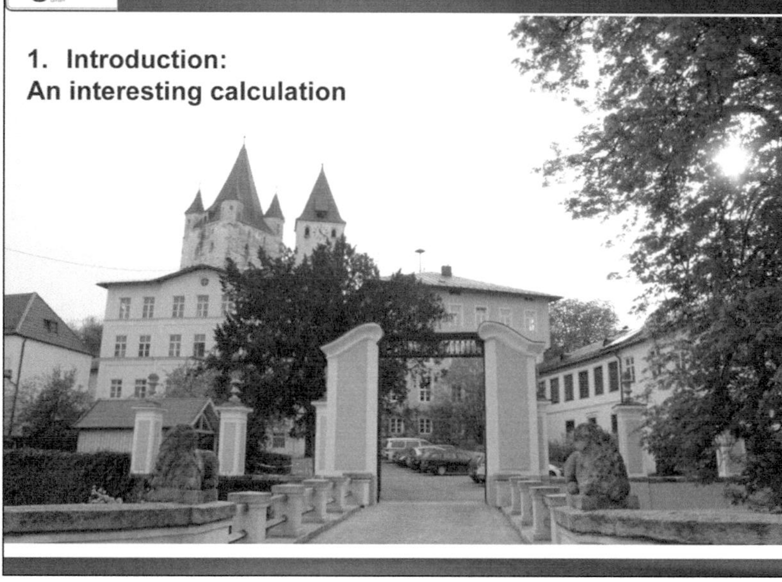

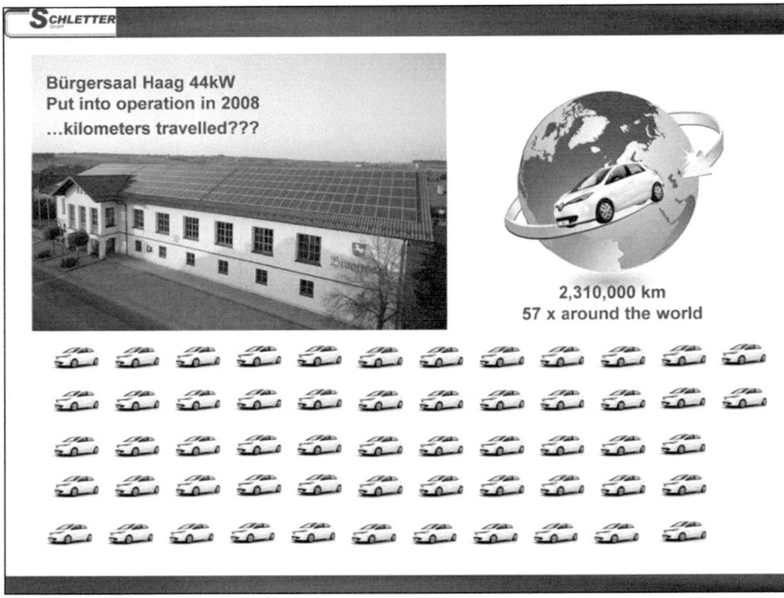
SCHLETTER
Bürgersaal Haag 44kW
Put into operation in 2008
...kilometers travelled???
2,310,000 km
57 x around the world

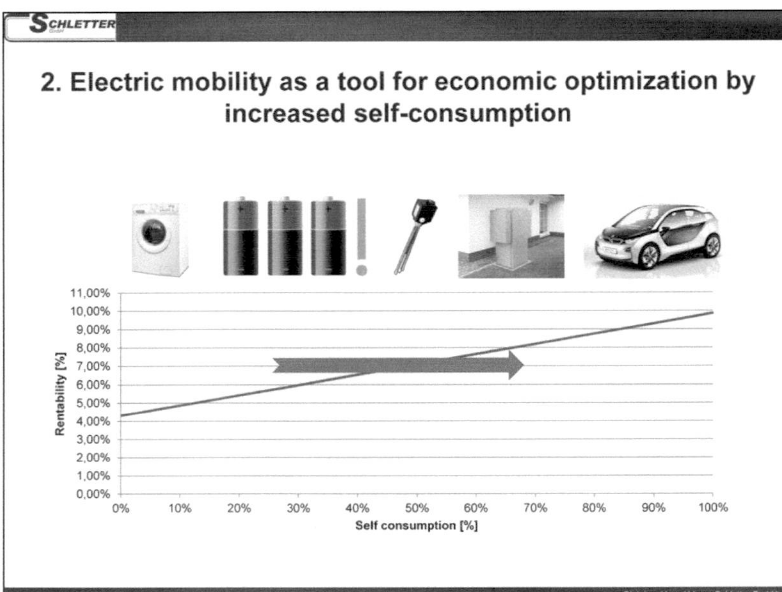
SCHLETTER
2. Electric mobility as a tool for economic optimization by increased self-consumption
Rentability [%]
11,00%
10,00%
9,00%
8,00%
7,00%
6,00%
5,00%
4,00%
3,00%
2,00%
1,00%
0,00%
0%
10%
20%
30%
40%
50%
60%
70%
80%
90%
100%
Self consumption [%]
Dipl. Ing. Hans Urban Schletter GmbH

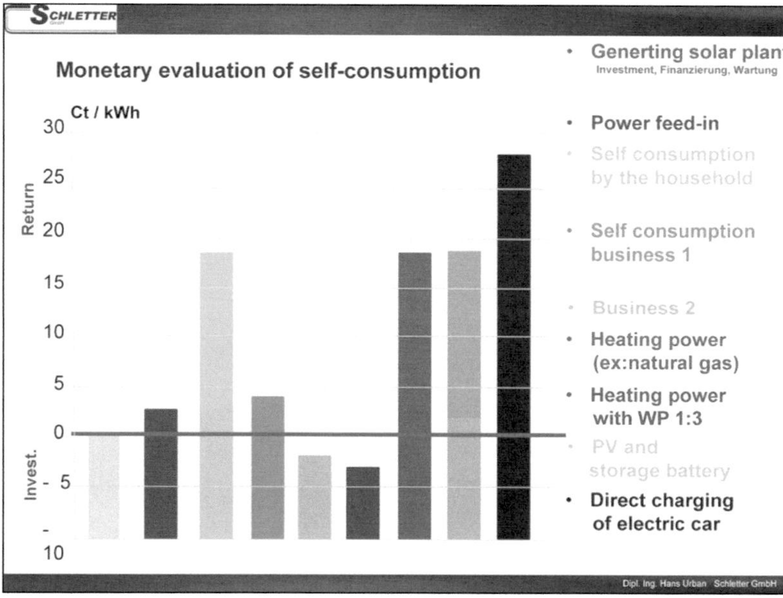
SCHLETTER
Monetary evaluation of self-consumption
Ct / kWh
30
25
20
15
10
5
0
- 5
-
10
Return
Invest.
Generting solar plant
Investment, Finanzierung, Wartung
Power feed-in
Self consumption by the household
Self consumption business 1
Business 2
Heating power (ex:natural gas)
Heating power with WP 1:3
PV and storage battery
Direct charging of electric car
Dipl. Ing. Hans Urban Schletter GmbH

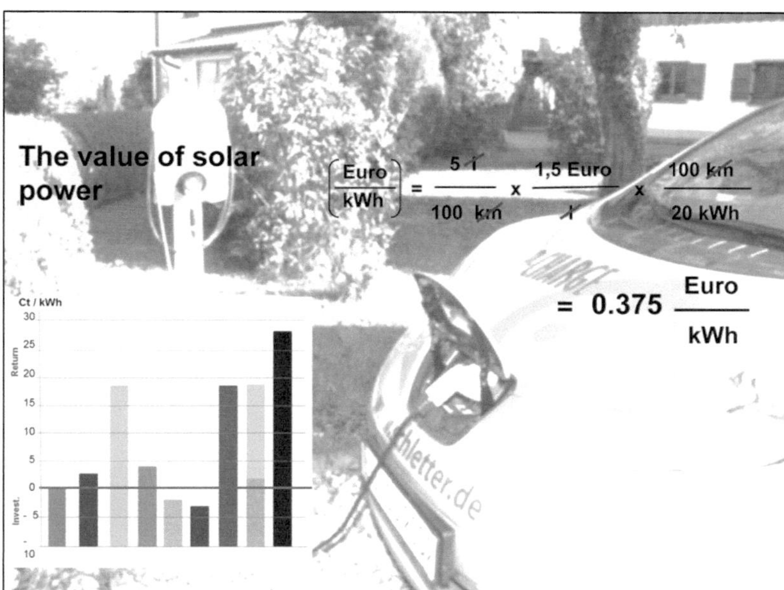
The value of solar power
(Euro / kWh) = 5 l / 100 km x 1,5 Euro / l x 100 km / 20 kWh
= 0.375 Euro / kWh
Ct / kWh
Return
Invest.
schletter.de

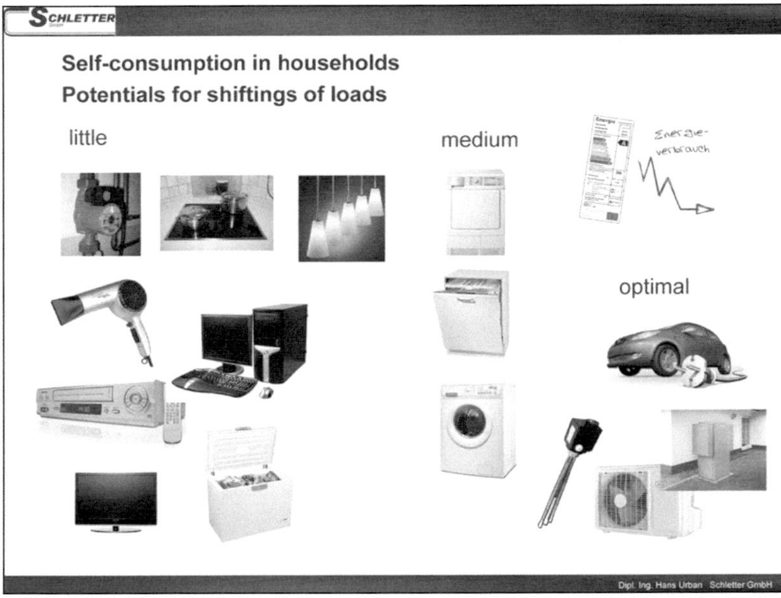
SCHLETTER
Self-consumption in households
Potentials for shiftings of loads
little
medium
Energie-
verbrauch
optimal
Dipl. Ing. Hans Urban Schletter GmbH

SCHLETTER
3. Sample 2012 – household without storage
Dipl. Ing. Hans Urban Schletter GmbH

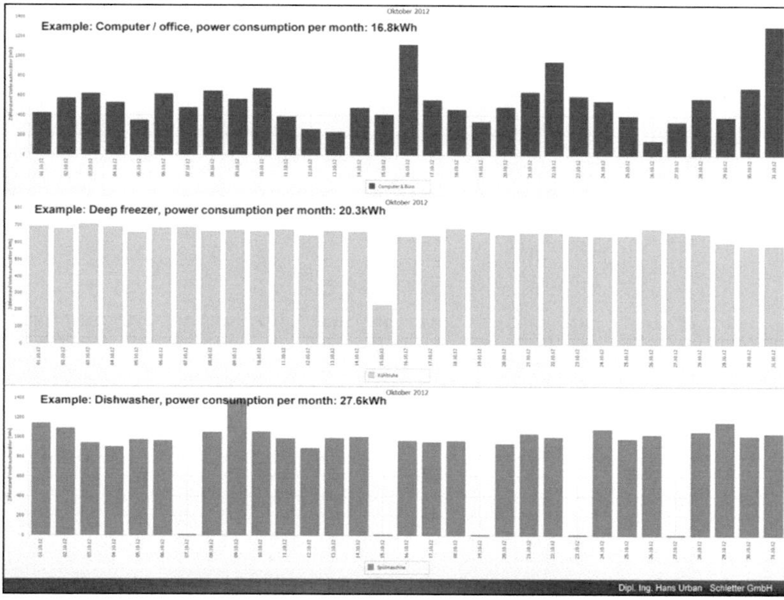
Oktober 2012
Example: Computer / office, power consumption per month: 16.8kWh
Oktober 2012
Example: Deep freezer, power consumption per month: 20.3kWh
Oktober 2012
Example: Dishwasher, power consumption per month: 27.6kWh
Dipl. Ing. Hans Urban Schletter GmbH

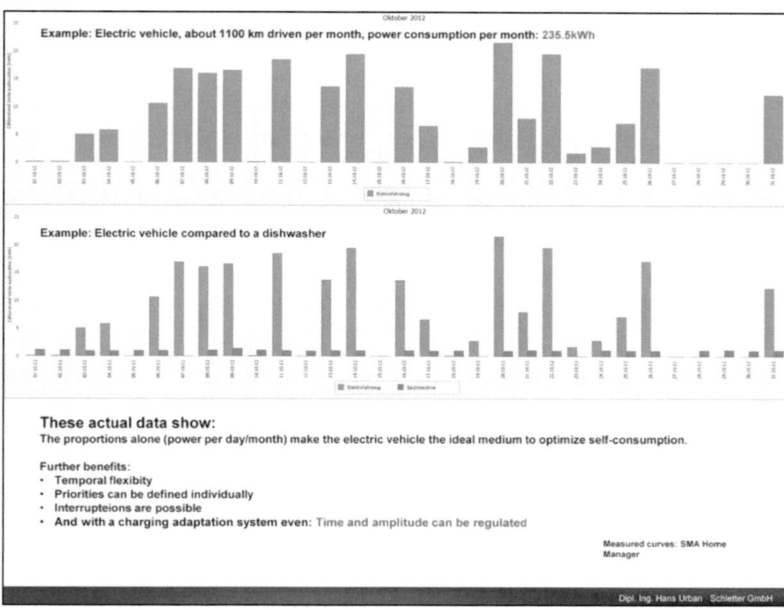
Oktober 2012
Example: Electric vehicle, about 1100 km driven per month, power consumption per month: 235.5kWh
Oktober 2012
Example: Electric vehicle compared to a dishwasher
These actual data show:
The proportions alone (power per day/month) make the electric vehicle the ideal medium to optimize self-consumption.
Further benefits:
• Temporal flexibity
• Priorities can be defined individually
• Interrupteions are possible
• And with a charging adaptation system even: Time and amplitude can be regulated
Measured curves: SMA Home Manager
Dipl. Ing. Hans Urban Schletter GmbH

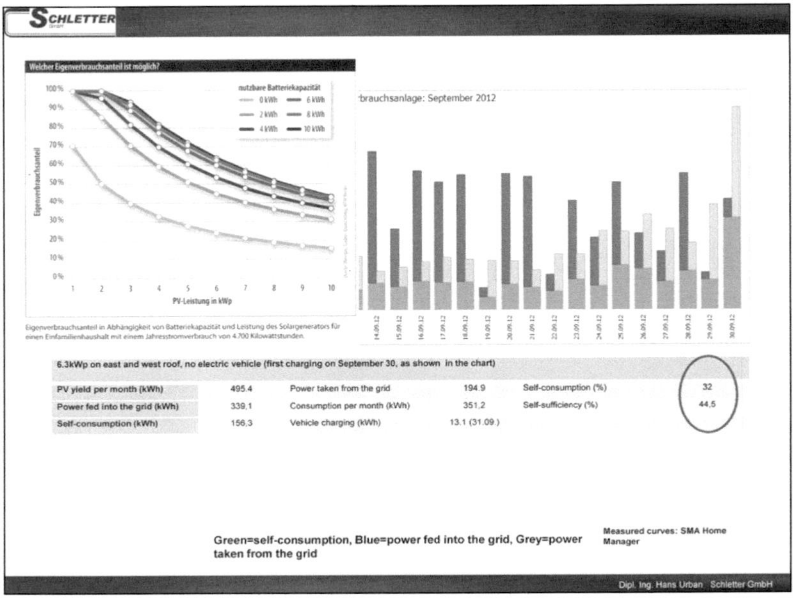
SCHLETTER
Welcher Eigenverbrauchsanteil ist möglich?
nutzbare Batteriekapazität
0 kWh
2 kWh
4 kWh
6 kWh
8 kWh
10 kWh
Eigenverbrauchsanteil
PV-Leistung in kWp
Eigenverbrauchsanteil in Abhängigkeit von Batteriekapazität und Leistung des Solargenerators für einen Einfamilienhaushalt mit einem Jahresstromverbrauch von 4.700 Kilowattstunden.
brauchsanlage: September 2012
6.3kWp on east and west roof, no electric vehicle (first charging on September 30, as shown in the chart)
PV yield per month (kWh) 495.4
Power fed into the grid (kWh) 339,1
Self-consumption (kWh) 156,3
Power taken from the grid 194.9
Consumption per month (kWh) 351,2
Vehicle charging (kWh) 13.1 (31.09.)
Self-consumption (%) 32
Self-sufficiency (%) 44,5
Green=self-consumption, Blue=power fed into the grid, Grey=power taken from the grid
Measured curves: SMA Home Manager
Dipl. Ing. Hans Urban Schletter GmbH

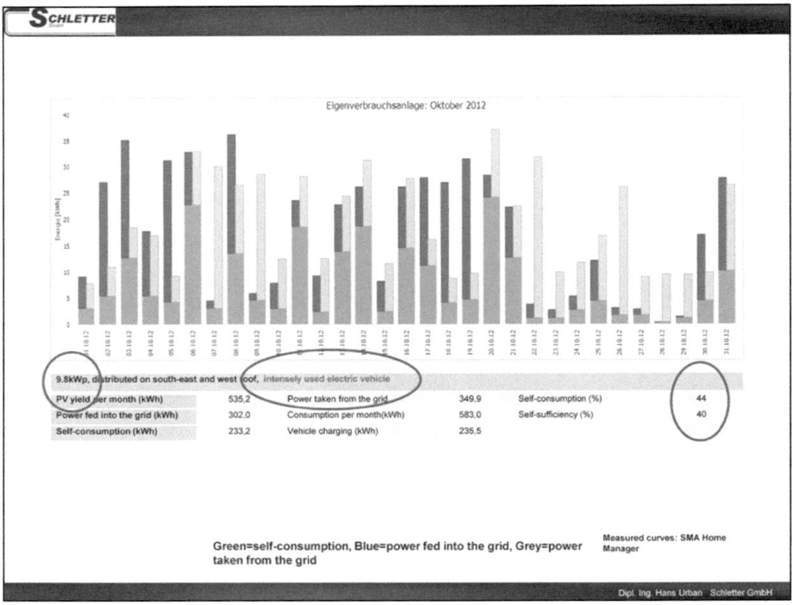
SCHLETTER
Eigenverbrauchsanlage: Oktober 2012
Energie [kWh]
9.8kWp, distributed on south-east and west roof, intensely used electric vehicle
PV yield per month (kWh) 535,2
Power fed into the grid (kWh) 302,0
Self-consumption (kWh) 233,2
Power taken from the grid 349,9
Consumption per month(kWh) 583,0
Vehicle charging (kWh) 235,5
Self-consumption (%) 44
Self-sufficiency (%) 40
Green=self-consumption, Blue=power fed into the grid, Grey=power taken from the grid
Measured curves: SMA Home Manager
Dipl. Ing. Hans Urban Schletter GmbH

SCHLETTER

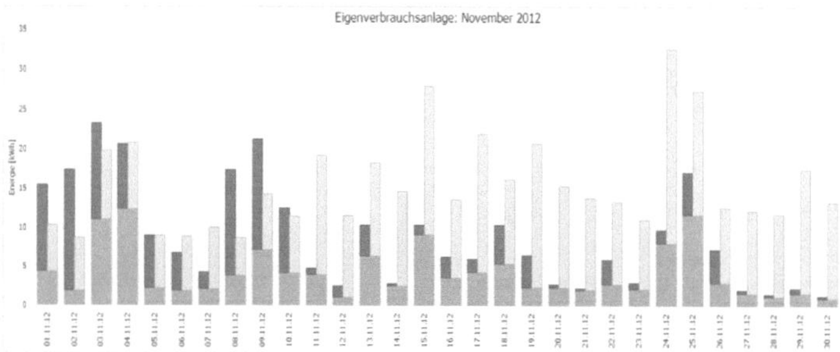

9.8kWp, arranged on south-east and west roof, electric vehicle charged every once in a while					
PV yield per month (kWh)	261,1	Power taken from the grid	341,7	Self-consumption (%)	48
Power fed into the grid (kWh)	136,3	Consumption per month (kWh)	466,4	Self-sufficiency (%)	26,7
Self-consumption (kWh)	124,7	Vehicle charging (kWh)	76,4		

Green=self-consumption, Blue=power fed into the grid, Grey=power taken from the grid

Messkurven: SMA Home Manager

Dipl. Ing. Hans Urban Schletter GmbH

SCHLETTER

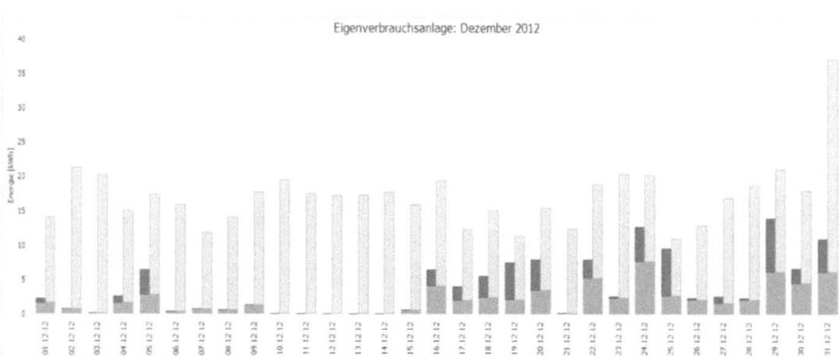

9.8kWp, arranged in south-east and west roof, electric vehicle almost never charged in the household					
PV yield per month (kWh)	117,9	Power taken from the grid	469,9	Self consumption (%)	55
Power fed into the grid (kWh)	53,3	Consumption per month (kWh)	534,5	Self-sufficiency (%)	12
Self-consumption (kWh)	64,6	Charging of vehicle (kWh)	17,2		

Green=self-consumption, Blue=power fed into the grid, Grey=power taken from the grid

Measured curves: SMA Home Manager

Dipl. Ing. Hans Urban Schletter GmbH

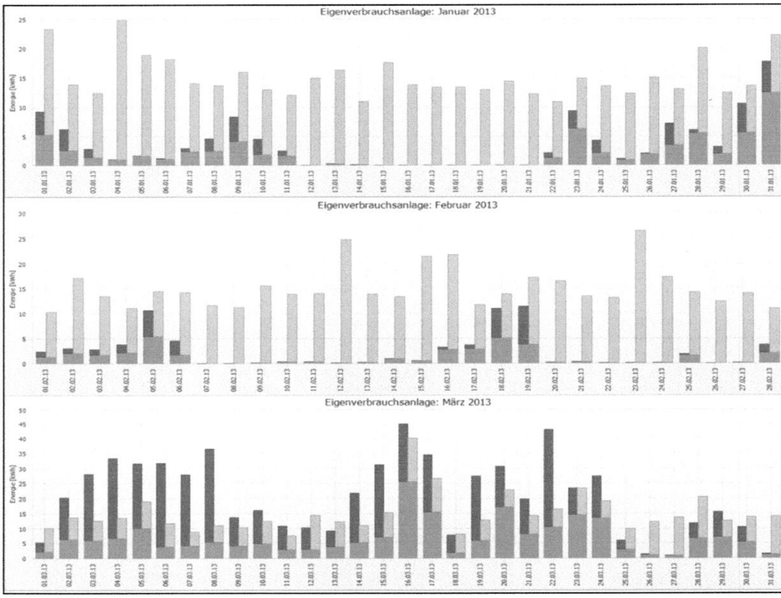
Eigenverbrauchsanlage: Januar 2013
Eigenverbrauchsanlage: Februar 2013
Eigenverbrauchsanlage: März 2013
Energie [kWh]

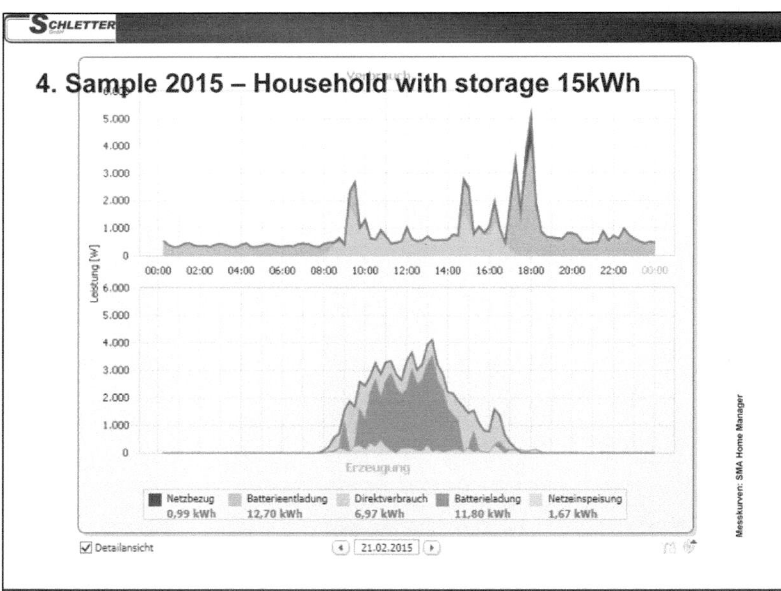
SCHLETTER
4. Sample 2015 – Household with storage 15kWh
5.000
4.000
3.000
2.000
1.000
0
00:00 02:00 04:00 06:00 08:00 10:00 12:00 14:00 16:00 18:00 20:00 22:00
Leistung [W]
6.000
5.000
4.000
3.000
2.000
1.000
0
Erzeugung
Netzbezug 0,99 kWh
Batterieentladung 12,70 kWh
Direktverbrauch 6,97 kWh
Batterieladung 11,80 kWh
Netzeinspeisung 1,67 kWh
Detailansicht
21.02.2015
Messkurven: SMA Home Manager
Dipl. Ing. Hans Urban Schletter GmbH

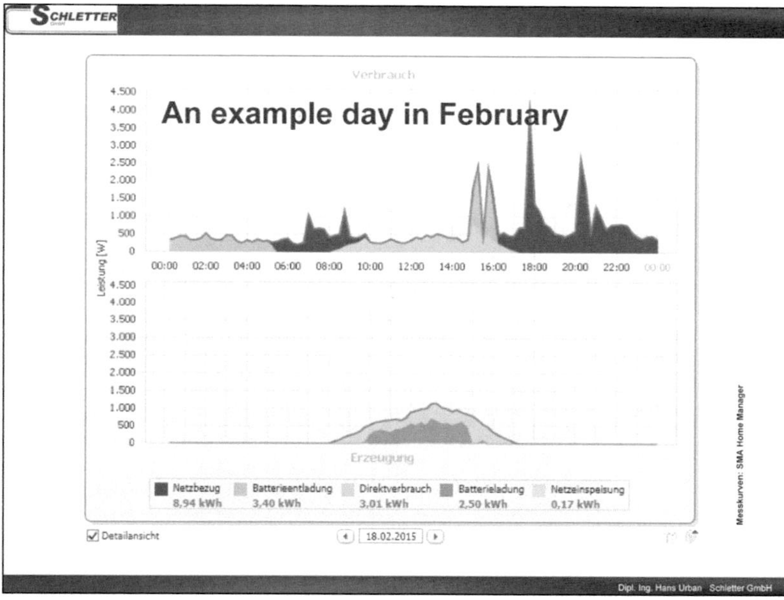
SCHLETTER
Verbrauch
An example day in February
4.500
4.000
3.500
3.000
2.500
2.000
1.500
1.000
500
0
Leistung [W]
00:00 02:00 04:00 06:00 08:00 10:00 12:00 14:00 16:00 18:00 20:00 22:00
Erzeugung
Netzbezug 8,94 kWh
Batterieentladung 3,40 kWh
Direktverbrauch 3,01 kWh
Batterieladung 2,50 kWh
Netzeinspeisung 0,17 kWh
Detailansicht
18.02.2015
Messkurven: SMA Home Manager
Dipl. Ing. Hans Urban Schletter GmbH

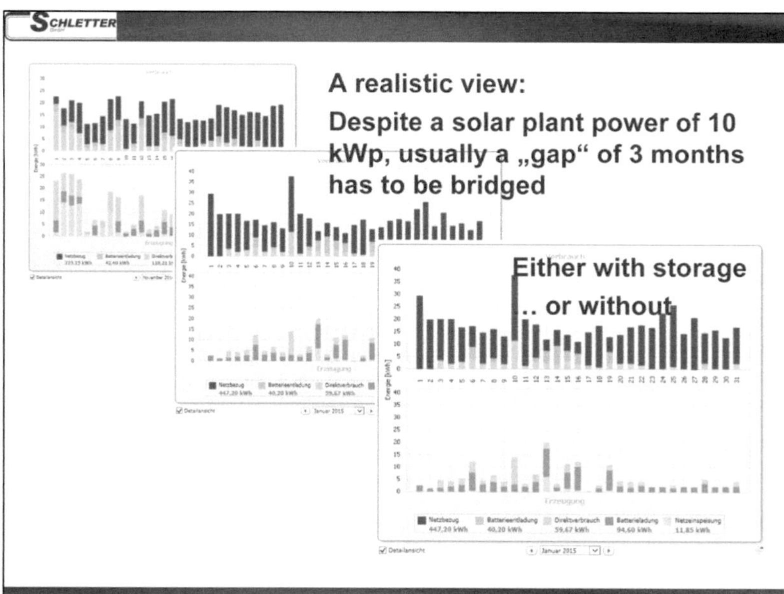
SCHLETTER
A realistic view:
Despite a solar plant power of 10 kWp, usually a „gap“ of 3 months has to be bridged
Either with storage
.. or without
Dipl. Ing. Hans Urban Schletter GmbH

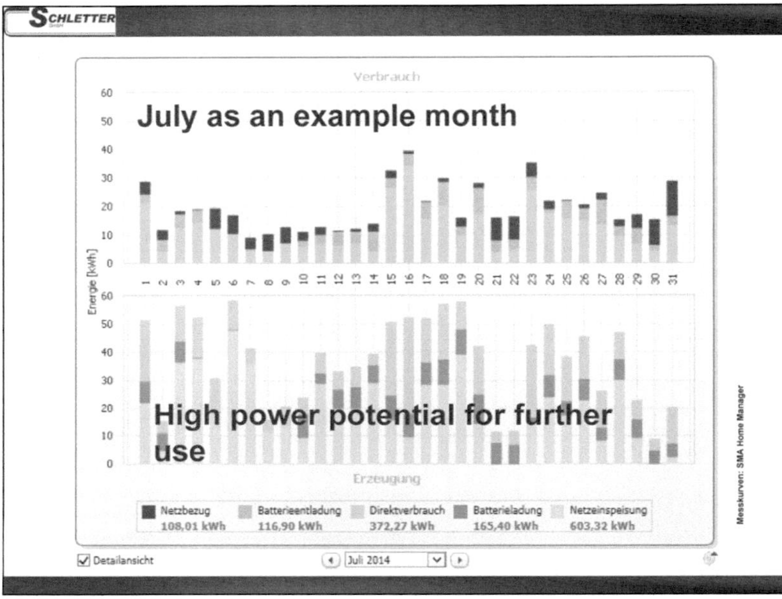
SCHLETTER
Verbrauch
July as an example month
Energie [kWh]
High power potential for further use
Erzeugung
Netzbezug 108,01 kWh
Batterieentladung 116,90 kWh
Direktverbrauch 372,27 kWh
Batterieladung 165,40 kWh
Netzeinspeisung 603,32 kWh
Detailansicht
Juli 2014
Messkurven: SMA Home Manager

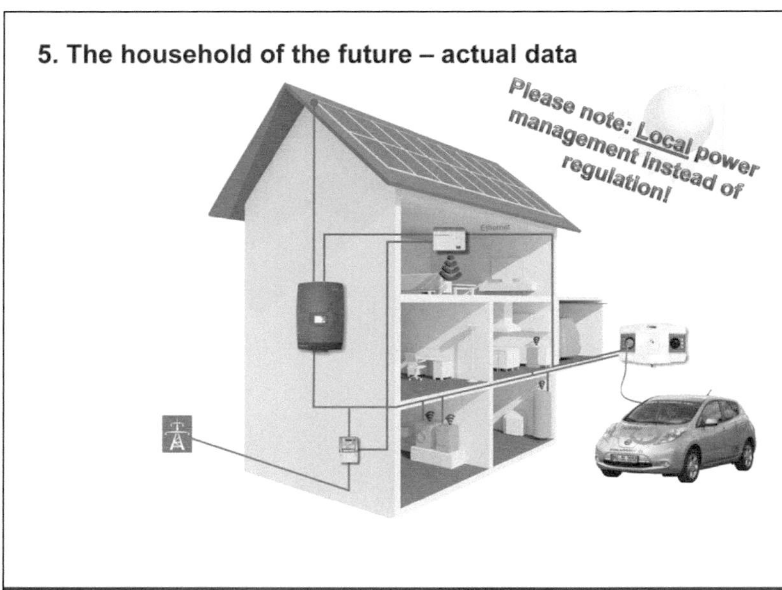
5. The household of the future – actual data
Please note: Local power management instead of regulation!

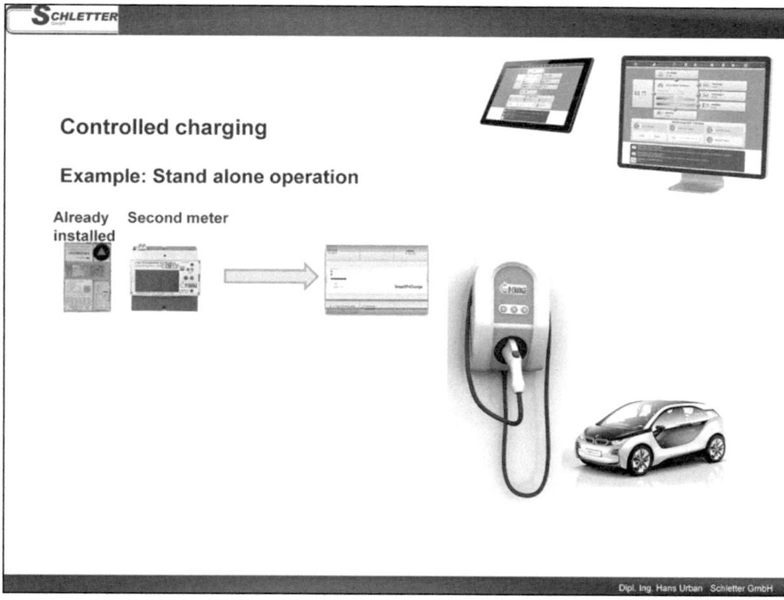
SCHLETTER
Controlled charging
Example: Stand alone operation
Already installed
Second meter
Dipl. Ing. Hans Urban Schletter GmbH

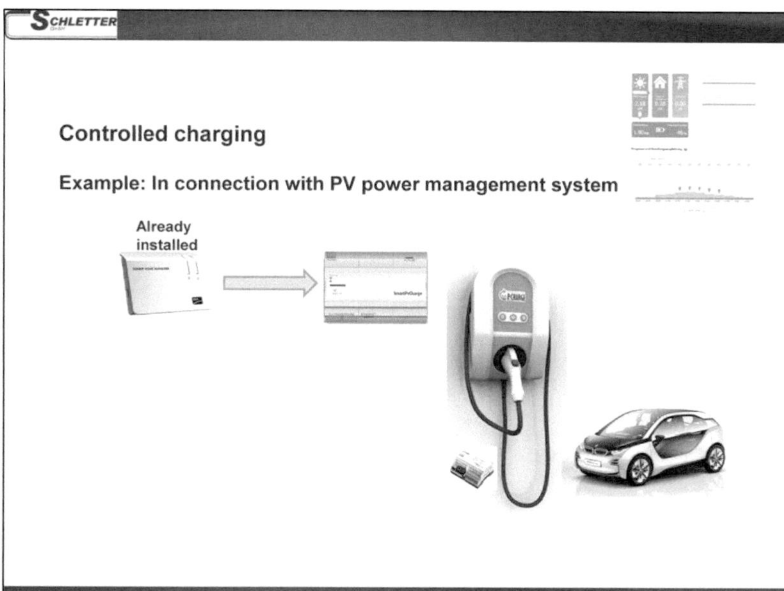
SCHLETTER
Controlled charging
Example: In connection with PV power management system
Already installed
Dipl. Ing. Hans Urban Schletter GmbH

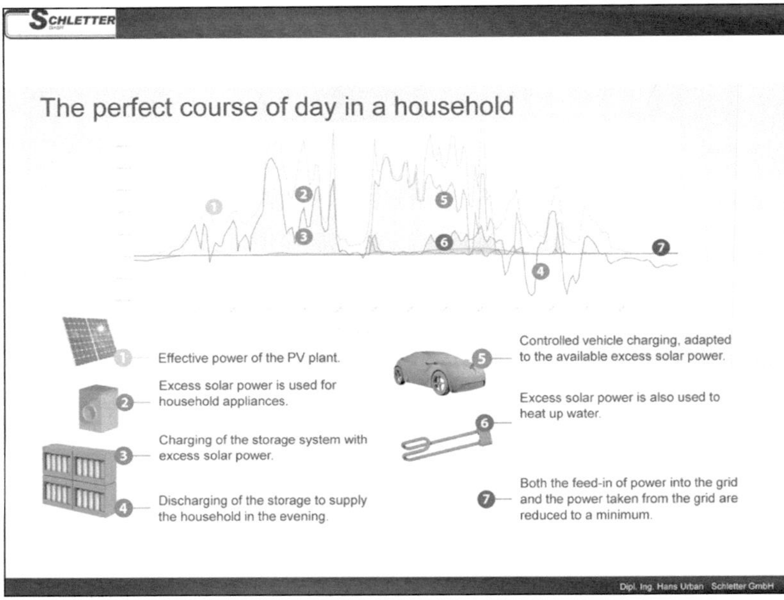
SCHLETTER
The perfect course of day in a household
Effective power of the PV plant.
Excess solar power is used for household appliances.
Charging of the storage system with excess solar power.
Discharging of the storage to supply the household in the evening.
Controlled vehicle charging, adapted to the available excess solar power.
Excess solar power is also used to heat up water.
Both the feed-in of power into the grid and the power taken from the grid are reduced to a minimum.
Dipl. Ing. Hans Urban Schletter GmbH

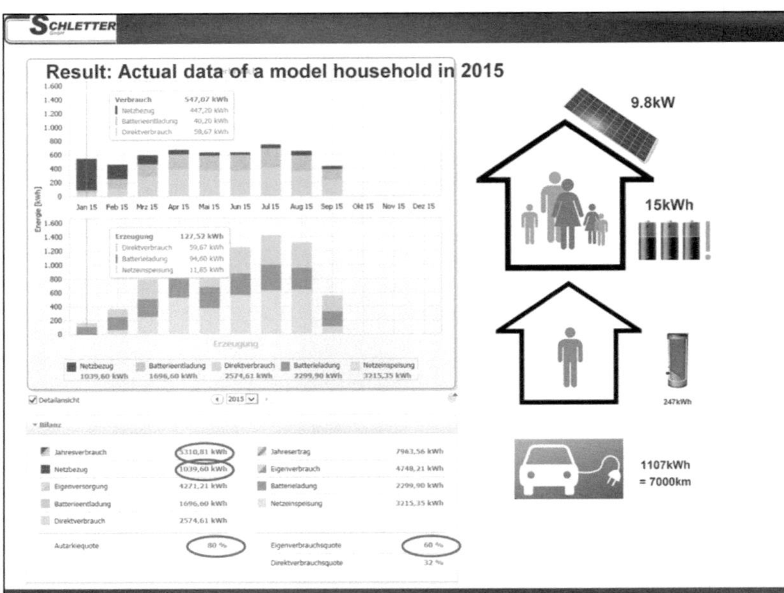
SCHLETTER
Result: Actual data of a model household in 2015
Netzbezug 1039,60 kWh
Batterieentladung 1696,60 kWh
Direktverbrauch 2574,61 kWh
Batterieladung 2299,90 kWh
Netzeinspeisung 3215,35 kWh
Detailansicht
2015
Bilanz
Jahresverbrauch 5310,81 kWh
Netzbezug 1039,60 kWh
Eigenversorgung 4271,21 kWh
Batterieentladung 1696,60 kWh
Direktverbrauch 2574,61 kWh
Jahresertrag 7963,56 kWh
Eigenverbrauch 4748,21 kWh
Batterieladung 2299,90 kWh
Netzeinspeisung 3215,35 kWh
Autarkiequote 80 %
Eigenverbrauchsquote 60 %
Direktverbrauchsquote 32 %
9.8kW
15kWh
247kWh
1107kWh
= 7000km

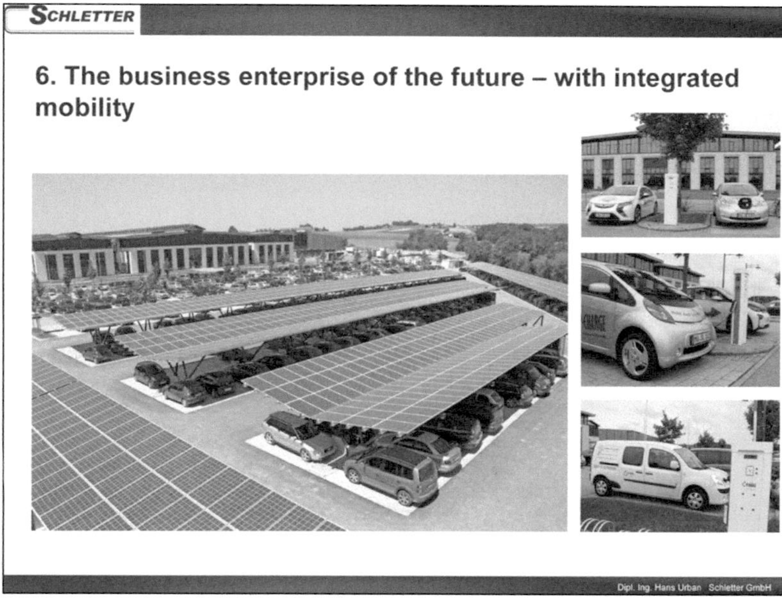
SCHLETTER
6. The business enterprise of the future – with integrated mobility
Dipl. Ing. Hans Urban Schletter GmbH

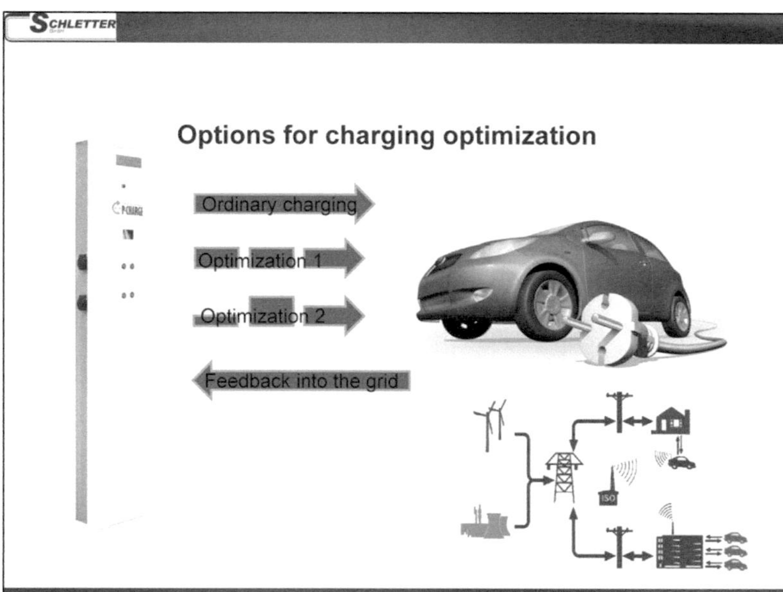
SCHLETTER
Options for charging optimization
Ordinary charging
Optimization 1
Optimization 2
Feedback into the grid
ISO
Dipl. Ing. Hans Urban Schletter GmbH

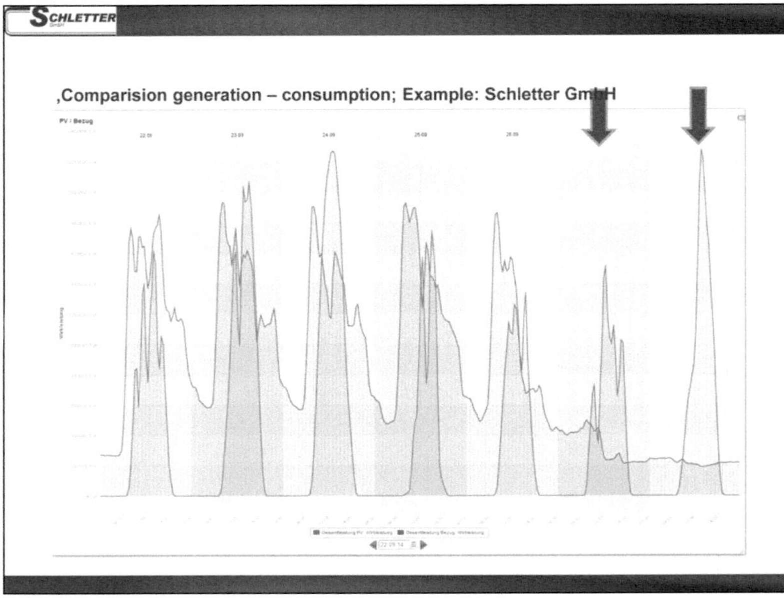
SCHLETTER
,Comparision generation – consumption; Example: Schletter GmbH
PV / Bezug

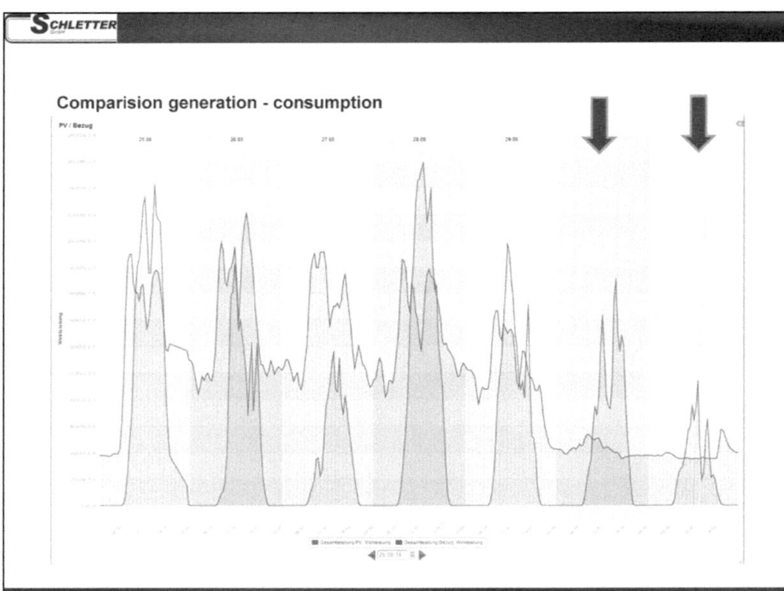
SCHLETTER
Comparision generation - consumption
PV / Bezug

7. PV carports as an ideal completion of the overall concept

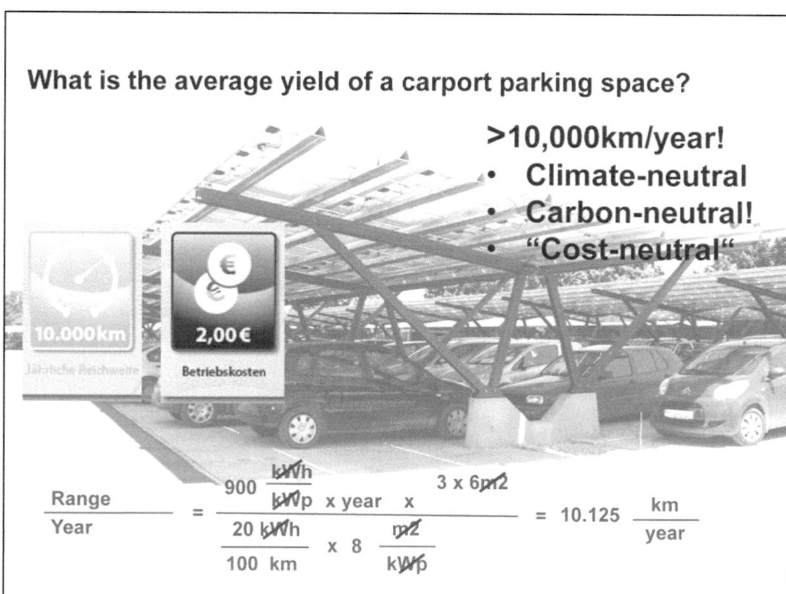
What is the average yield of a carport parking space?
>10,000km/year!
• Climate-neutral
• Carbon-neutral!
• "Cost-neutral"
10.000km
Jährliche Reichweite
2,00€
Betriebskosten
Range / Year = (900 kWh/kWp x year x 3 x 6m2) / (20 kWh/100 km x 8 m2/kWp) = 10.125 km/year

SCHLETTER
8. PV carports: Examples
• ca. 150 kWp - 2012
Dipl. Ing. Hans Urban Schletter GmbH

SCHLETTER
• 1.7 MWp - 2012
Dipl. Ing. Hans Urban Schletter GmbH

SCHLETTER

• 52 kWp - 2010

Dipl. Ing. Hans Urban Schletter GmbH

SCHLETTER

• 1 MWp - 2011

Dipl. Ing. Hans Urban Schletter GmbH

SCHLETTER
• 600 kWp - 2012
Dipl. Ing. Hans Urban Schletter GmbH

SCHLETTER
• 1 MWp - 2010
Dipl. Ing. Hans Urban Schletter GmbH

SCHLETTER
• 5.9 MWp - 2010
Dipl. Ing. Hans Urban Schletter GmbH

SCHLETTER
• 613 kWp - 2010
Dipl. Ing. Hans Urban Schletter GmbH

SCHLETTER
• 15 kWp - 2011
Dipl. Ing. Hans Urban Schletter GmbH

SCHLETTER
• 288 kWp - 2013
Dipl. Ing. Hans Urban Schletter GmbH

SCHLETTER

• 500 kWp - 2010

Dipl. Ing. Hans Urban Schletter GmbH

SCHLETTER

... even on the "top of the world"

Thank you for your attention!

Dipl. Ing. Hans Urban Schletter GmbH

INFORMATION AND COMMUNICATION TECHNOLOGY

Data communication and grid quality on charging interface

Dipl.-Ing. (BA) Ursel Willrett

IAV GmbH, Sindelfingen, Germany

1 Introduction

1.1 Motivation

The charging interface between grid and electric vehicles is new. Energy and automotive industry have to co-operate for a successful introduction of e-mobility. There are plenty of challenges to provide a proper co-operation of these "two worlds". From the view of the energy provider integration of e-mobility into smart grid is important, which includes effective dynamic load management and the use of the HV-Batteries in the vehicles to store and feedback energy. The user expects sufficient and reliable charging points to re-charge his electric vehicle everywhere and at any time. Keys for an infrastructure accepted by users are easy to use and secure data transfer of personal data.

Therefore important challenges for introduction of e-mobility are:

- Communication procedures and data security in the whole system
- Grid stability
- Dynamic load management

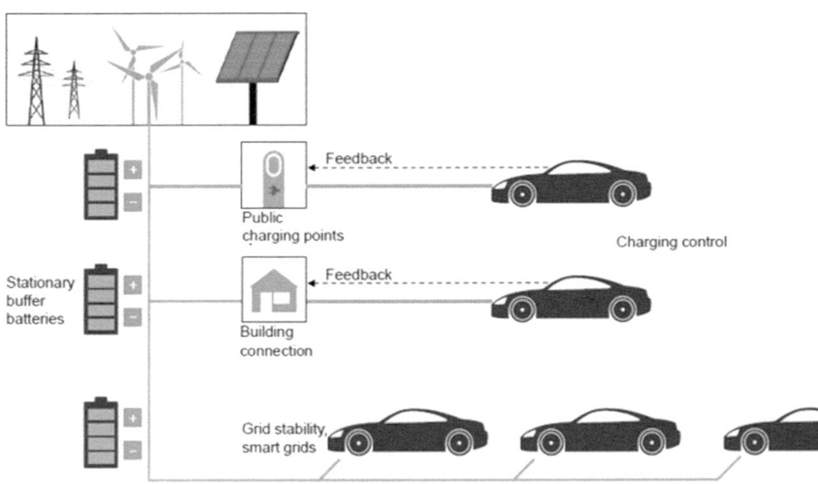

Network integration of electric vehicles [IAV]

The following chapters describe secured data communication and grid stability in more detail.

1.2 Requirements for integration into grid

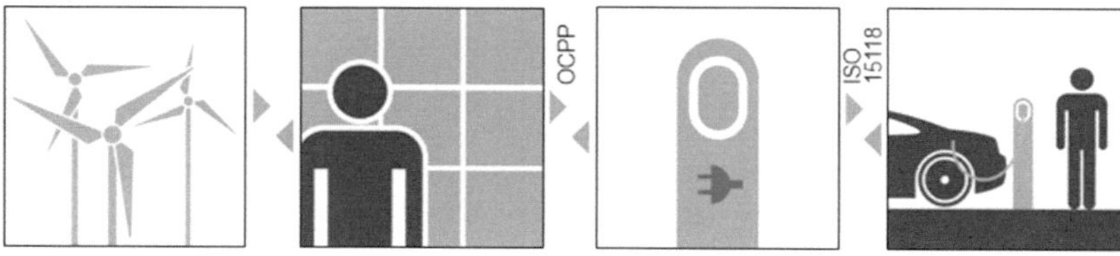

Primary actors and secondary actors [IAV]

The system consists of many parties. The electric vehicles and charging stations are primary actors [2]. All parties beyond the charging station from the view of the user of an electric vehicle are called "secondary actors" [2]. All components are connected via a communication network and full-fill the communication standards to exchange all necessary information to provide the desired functions.

Some of the functions performed by the network:

Group	Applications
Authorization	• Authentication at any charging station consistently and independent of the operator of the infrastructure
Charging Control	• Charging control controlled by voltage or current • Similar communication procedures independent on the current type (alternating current AC or direct current DC)
Grid stability	• Reduced noise and distortion caused by electric vehicles using adequate filters to comply with the standards • Robustness in the electric vehicles to resist noise and distortion in the grid
Load Management	• Battery optimized charging (charging power, temperature, State of Charge) and cost controlled charging (i.e. beneficial tariffs) • Smart-Grid options for optimized use of renewable energies • Fleet management with a huge number of electric vehicles (Drive & Charge)
Billing	• Automatic payment • Access to all charging station with one unique contract (e-roaming) • Data security using professional processes with signatures and certificates
Value added services	• Electric vehicle status information (i.e. State of Charge, remaining charging time) • Time controlled charging (i.e. input of planned departure) • Access to services via internet, i.e. software updates, diagnostics

1.3 Load management

Load management functions are performed to balance the available energy and the demand of energy by the users (smart grid). Charging requirements by the users are classified into three groups:

- Charging on demand (stochastic): the user expects to receive energy immediately
- Controlled charging: the user expects a fully charged battery at desired time (driven by financial reward)
- Dynamic load management: use of the electric vehicles as electricity storage system. This option includes grid feedback facilities.

Requirement for load management – example:

Several vehicles intend to charge at the same time on 3-phase 63 A, with 43,5 kW. The local grid provides not more than 680 kVA. The grid is overloaded from as few as 16 electric vehicles.

To avoid overloading the grid by means of intelligent control (load management) it is required to provide a communication protocol stack. The ISO/IEC 15118 specifies the appropriate control commands for charging communication. On the power lines the limit values must be met for maximum grid distortion (DIN EN 50160 and IEC 61000).

Secured data communication and grid stability are key to perform the necessary functions for load management.

2 Data communication

2.1 Communication between electric vehicle, charging station and backend

ISO/IEC 15118 is the specified protocol stack on Power Line Communication (PLC) between electric vehicles and charging station [3] modulated on a basic pulse-width communication (PWM) [1]. ISO/IEC 15118 describes all sequences, messages and parameters to provide the required applications. Data security methods are specified, i.e. encryption methods, formats and contents of certificates.

Communication protocols electric vehicle – charging station – backend [IAV]

For data which are transferred between electric vehicles and the backend (secondary actors) the charging station acts as a gateway. Between charging station and the backend adequate protocols are applicable (i.e. OCPP [4]). Several protocol stacks are currently used, a world-wide standard is actually developed.

2.2 Data security

2.2.1 General requirements

Data security covers three main properties:

Authenticity: The communication parties are really those which they claim to be.

Confidentiality: The contents of a message can be read only by intended recipients, not by unauthorized third parties.

Integrity: Unauthorized modification of the sent message must be avoided or at least detected.

2.2.2 Data security methods standardized in ISO/IEC 15118

The protocol stack of the ISO/IEC 15118 [3] provides encryption in two of the layers. In layer four the transport layer security protocol (TLS) is specified. TLS supports encryption of the data between electric vehicle and charging station. In layer seven XML security is available with encryption of data between electric vehicle and secondary actors. Asymmetric and symmetric encryption algorithms are used for secure data transmission.

Encryption covers confidentiality of the two communication entities. It is a two stage concept: lower layer encryption between electric vehicle and charging station and higher layer encryption between electric vehicle and secondary actor (end-to-end encryption). It is therefore not possible to read personal data in the charging station.

The methods for handling of certificates and signatures including the respective data formats are also described in ISO/IEC 15118. Creation and verification of digital XML based signatures is the method to cover authenticity and integrity. In the ISO/IEC 15118 seven types for certificates used for different purposes which are handled by different instances are specified: V2G root certificate, charge point operator certificate, mobility operator root certificate, contract certificate, OEM root certificate, OEM provisioning certificate, private operate root certificate [3]. According to ISO 15118 the chain for the certificates consists of maximal three elements deviated by the root certificate. The lower layer certificates are signed by the higher layer certificate. For validation of authenticity of the certificates private and public keys are used.

2.2.3 Data to be protected by unauthorized access

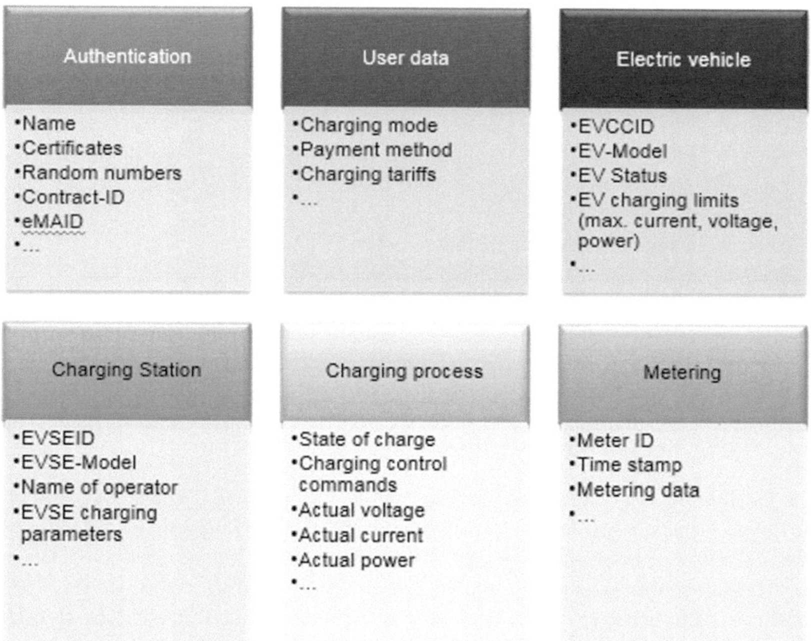

Data to be secured [IAV]

It seems to be a huge effort to provide data security. The data to be protected are personal data which the user does not want to be read by unauthorized persons. In addition further data (i.e. metering data) are transmitted which are used to create bills. They must not be read or manipulated by third parties. Otherwise the confidence in e-mobility is lost. In the figure above data to be secured are sorted in groups including a few examples for sensitive data.

2.2.4 Status and further challenges for implementation of data security

Methods are available to provide secured data communication between electric vehicle, charging station and all secondary actors in the backed. In the ISO/IEC 15118 the communication between electric vehicle and charging station is specified. The standard includes also the description of the adequate encryption methods and the handling of signatures and certificates. All specified methods comply with modern public key infrastructure systems.

The methods are specified, but the standards do not describe the process how to handle data security in the system. Because of the e-mobility system is built and maintained by several parties (i.e. automotive industry, charging station manufacturers, energy suppliers, providers, further secondary actors, root instances) the definition and establishment of a process is required with clearly defined roles. Some examples of topics are:

- World-wide standard of the communication between charging station and Backend
- Administration of the various certificates (i.e. OEM, charge point operator CPO, provider)
- Encryption key handling and updates
- Secured storage of private keys and certificates
- Administration of several contract, e-roaming

3 Grid quality

3.1 Risks caused by distortion

Electric consumers (i.e. electric vehicles) need on the one hand a stable power supply, they expect grid quality according to the given standards. Otherwise every consumer adds distortion or noise into the grid. Today only few electric vehicles exist. During the charging process of already one electric vehicle perturbations are visible. These non-linear distortions are caused mainly by semiconductors used in charging rectifiers either in the charging station (DC charging) or in electric vehicles (AC charging). Predictions of distortions for many electric vehicles cannot be determined, it is known that the results cannot be calculated by a linear approximation. Reductions of noise and distortions is performed by adequate filters and compensation circuits.

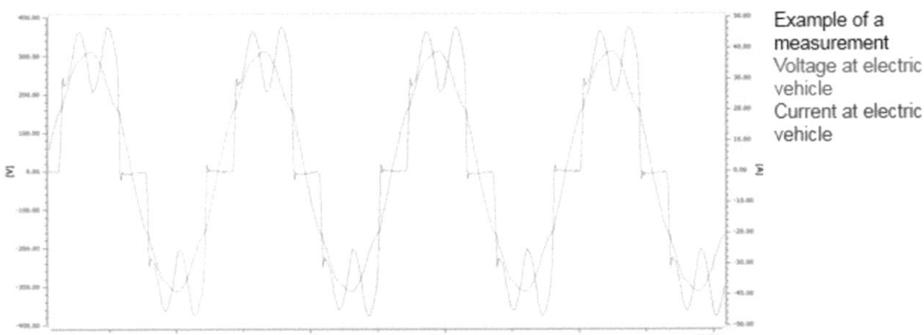

U(t) and i(t) during charging process of one electric vehicle [IAV]

Grid distortions may cause interruptions in power supply systems. Unexpected break-offs of the charging process may be caused by grid distortions. Devices and equipment connected at the same power distribution may be damaged or destroyed, fuses may fire.

3.2 Requirements and standardization

Voltage quality and electromagnetic compatibility (EMC) are linked. Definition of grid quality according to IEC 61000 ([6]-[11]):

„Characteristics of electricity at a certain location in an electrical system, the characteristics is rated based on a selection of technical reference parameters."

This definition is location independent and includes EMC compliance.

The characteristics for grid quality from the view of the power network is specified in EN 50160 [5]. The standard contains the tolerance thresholds for permitted distortions caused by connected consumers.

Examples for parameters according to EN 50160:

Total Harmonic Distortion (THD)

The total harmonic distortion describes the grade of distortion either on voltage or on current. It is the relation of the sum of harmonics related the basic oscillation.

$$\mathrm{THD}_I = \mathrm{d} = \frac{\sqrt{\sum_{n=2}^{x} I_n^2}}{I_1} = \frac{\sqrt{I^2 - I_1^2}}{I_1}$$

$$\mathrm{THD}_U = \mathrm{d} = \frac{\sqrt{\sum_{n=2}^{x} U_n^2}}{U_1} = \frac{\sqrt{U^2 - U_1^2}}{U_1}$$

The following figure shows the sine wave of an alternating current with a THD of 7%. The amplitude is attenuated by the high frequent harmonics.

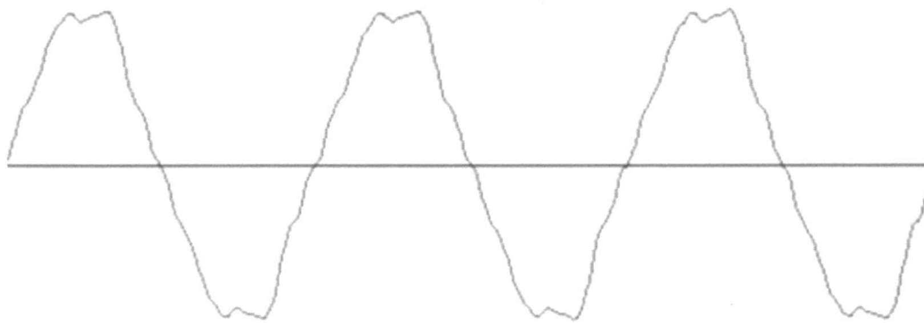

The consequences are additional loss, i.e. idle currents in transformers, lines and electronic devices containing capacitors or power chokes. Onboard chargers in electric vehicles are strongly affected.

Active factor or power factor ($\cos\varphi$)

The active factor $\cos\varphi$ is equal to the relation of the active power to the basic oscillation of the apparent power.

$$\cos\varphi = \frac{P}{|S|}$$

3.3 Measurement of grid quality

3.3.1 Network analyzer

The measurement of the parameters of the grid quality is performed by respective network analyzers. They record voltage and current of 1-phase or 3-phase connections configured by the desired measurement rate. From the raw current and voltage values all parameters specified in EN 50160 [5] are determined. Using evaluation software on the PC a full analysis is performed.

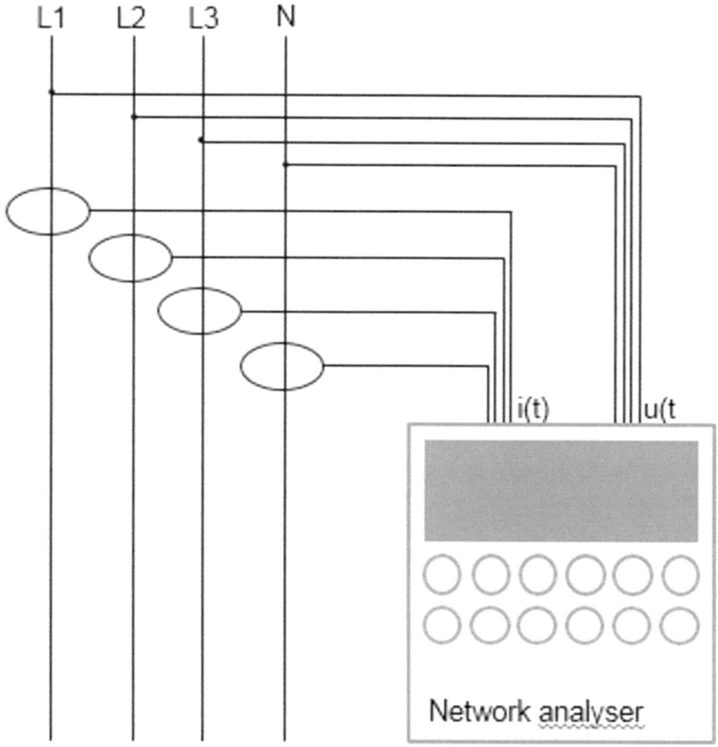

Measurement of grid quality by network analyser [IAV]

To access to the power lines on the charging interface adapters are required to allow the connection of the measurement lines to the three phases. For AC charging (charging mode 2 and 3) the measurement is performed between charging station or plug socket and electric vehicle. For DC charging (charging mode 4) the network analyzer is connected at the AC power supply connection of the charging station.

3.3.2 Measurement results – example

An electric vehicle had been measured at a public charging station with several connectors. It had been charged with alternating current, 1 phase at 230 V with max. 16A. A total harmonic distortion (THD) had been measured constantly of 2,6%, which is within the specified thresholds.

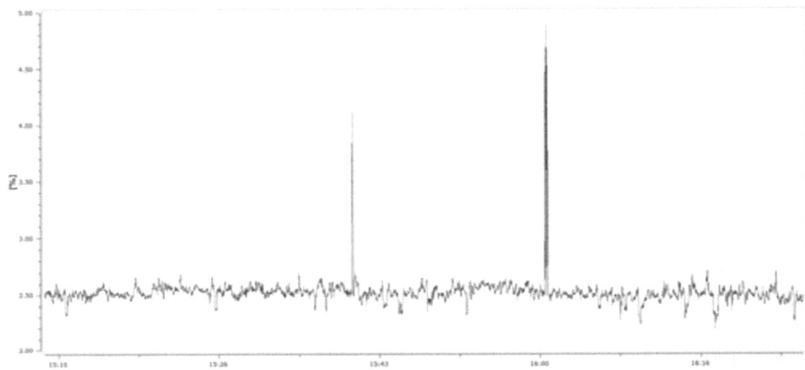

Measurement example: Total harmonic distortion – AC charging [IAV]

The figure above shows the measurement results over measurement time. Distortion peaks had been observed with doubling of THD to 5%. This occurred in the power connection during insertion of a plug of another electric vehicle. The drop of the THD down to 2,6% is performed by an active filter built in the onboard charger. The active filter reacts with a delay. The peak distortions cannot be avoided. During concurrent charging of several electric vehicles damages of connected electric equipment may be possible.

3.4 Methods to reduce distortion

Reduction of distortion and noise is performed using filters. There are circuit concepts which differentiate by complexity, quality and cost. Adequate circuit concepts are power chokes, passive filters and active filters (power factor correction filters PFC).

To design a proper concept the type of noise or distortions has to be identified. Many of them are not constant but change by time.

The most simple circuit concept is the use of power chokes which contain passive inductive components. They enhance current flow and decrease current peaks. Power chokes attenuate harmonics but they are not capable to filter or eliminate the harmonics.

Using passive filters harmonics can be filtered using a bandpass. A bandpass eliminates the respective harmonics at its resonance frequency. A bandpass is designed for a fixed resonance frequency. Changes in grid performance (i.e. dynamic change in the noise behavior) cannot be adapted dynamically.

Dynamic adaptation to noise behavior is performed using active filters. Usually power factor correction filters (PFC) are used. The charging current is controlled to provide an almost perfect sinus wave for the voltage at the power supply access. This circuit concept is also appropriate for bi-directional chargers (example for circuit structure s. figure below).

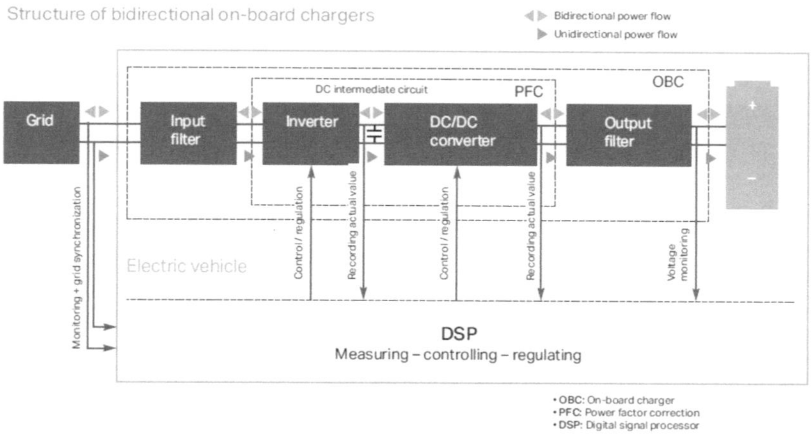

Bi-directional on-board charger – block diagram [IAV]

The control of the filter is performed by the digital signal processor. In the figures below an example for a filter with good compensation and an example for a filter with poor compensation is presented.

Good compensation – circuit simulation

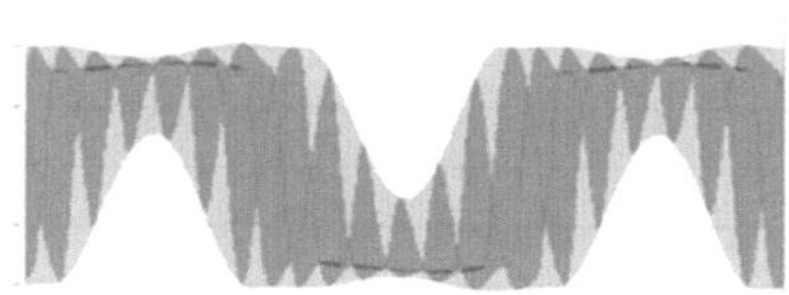

Poor compensation – simulation [IAV]

The system perturbations of a charging rectifier are definitely dependent on the design of PFC filter and its efficiency.

Predictions of distortions for many electric vehicles cannot be determined easily. The measured effects are not linear.

4 Conclusion

Acceptance of E-mobility by users is key for market success. E-mobility has to be easy to use and transparent. The users want to charge everywhere and expect security for their personal data.

The main technical challenges for introduction of E-mobility are

- Unique communication procedures and data security in the whole system
- Grid stability
- Dynamic load management

The standard ISO/IEC 15118 specifies the communication between electric vehicle and charging station. It also includes data security methods with encryption and handling of certificates and signatures. For integration into a consistent system concept adequate processes have to be specified, agreed and supported by all parties (automotive industry, energy providers, charging station manufacturers and users).

Providing grid quality is also essential to charge properly without unexpected breaks during the charging process. Distortions and noise on the charging interface has to be limited to avoid unintended incidents. Using powerful power factor correction filters in the AC/DC converters significant reduction of distortions and noise is feasible. Regular measurements to validate the distortions are necessary to prove that the noise keeps within the thresholds given by the standards. An overall prediction of the noise created by increasing number of electric vehicles at a charging point is not possible by a mathematical model. The measurements results will support further improvements in the circuits of the AC/DC converters covering the requirements with an increased number of electric vehicles.

5 Bibliography

1. International Electrotechnical Commission (2013): IEC 61851-1 – Electric vehicle conductive charging system – Part 1: General requirements

2. ISO 15118 (2013): Road vehicles – Vehicle to grid communication interface – Part 1: General Information and use case definition

3. ISO 15118 (2013): Road vehicles – Vehicle to grid communication interface – Part 2: Network and application protocol requirements

4. Open Charge Alliance (2014): Open Charge Point Protocol 2.0 – Interface description between Charge Point and Central System, URL: http://www.openchargealliance.org/sites/default/files/OCPP%202.0%20Release%20Candidate%202.pdf

5. DIN EN 50160 Merkmale der Spannung in öffentlichen Elektrizitätsversorgungsnetzen, Frankfurt: Beuth Verlag, 2011-02

6. DIN EN 61000-2-2 Elektromagnetische Verträglichkeit (EMV) Umgebungsbedingungen; Verträglichkeitspegel für niederfrequente leitungsgeführte Störgrößen und Signalübertragung in öffentlichen Niederspannungsnetzen, Frankfurt: Beuth Verlag, 2003-02.

7. DIN EN 61000-2-12 Elektromagnetische Verträglichkeit (EMV) – Teil 2-12: Umgebungsbedingungen – Verträglichkeitspegel für niederfrequente leitungsgeführte Störgrößen und Signalübertragung in öffentlichen Mittelspannungsnetzen, Frankfurt: Beuth Verlag, 2004-01

8. DIN EN 61000-3-2 Elektromagnetische Verträglichkeit (EMV) Grenzwerte – Grenzwerte für Oberschwingungsströme (Geräte-Eingangsstrom <= 16 A je Leiter), Frankfurt: Beuth Verlag, 2010-03

9. DIN EN 61000-3-3 Elektromagnetische Verträglichkeit (EMV) -Teil 3-3: Grenzwerte -Begrenzung von Spannungsänderungen, Spannungsschwankungen und Flicker in öffentlichen Niederspannungs-Versorgungsnetzen für Geräte mit einem Bemessungsstrom <= 16 A je Leiter, Frankfurt: Beuth Verlag, 2012-07

10. DIN EN 61000-3-12 Elektromagnetische Verträglichkeit (EMV): Grenzwerte - Grenzwerte für Oberschwingungsströme, verursacht von Geräten und Einrichtungen mit einem Eingangsstrom > 16A und <= 75A je Leiter, die zum Anschluss an öffentliche Niederspannungsnetz, Frankfurt: Beuth Verlag, 2011-03

11. DIN EN 61000-4-4 Elektromagnetische Verträglichkeit (EMV) – Teil 4-4: Prüf- und Messverfahren – Prüfung der Störfestigkeit gegen schnelle transiente elektrische Störgrößen/Burst, Frankfurt: Beuth Verlag, 2010-11

ISO 15118 – charging communication between plug-in electric vehicles and charging infrastructure

Dr. Andreas Heinrich,
Daimler AG

Michael Schwaiger,
BMW Group

DAIMLER
ISO 15118: Charging Communication Between Plug-In Electric Vehicles and Charging Infrastructure
ATZ Conference: Grid Integration of Electric Mobility
Andreas Heinrich, 2016-06-01

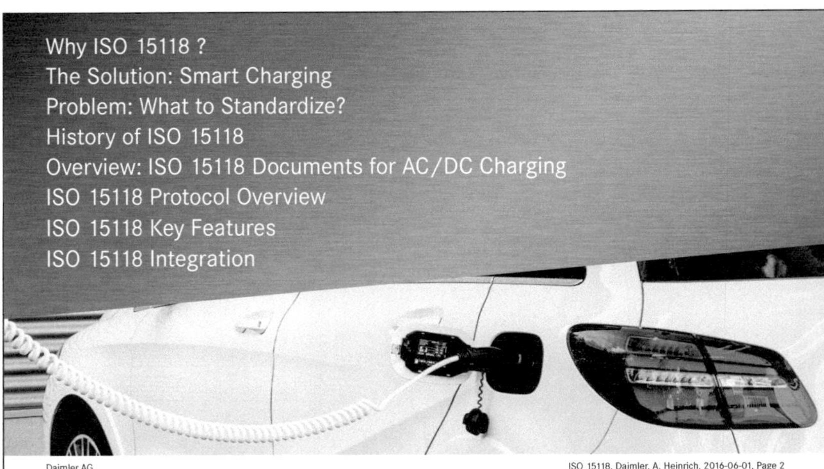
Why ISO 15118 ?
The Solution: Smart Charging
Problem: What to Standardize?
History of ISO 15118
Overview: ISO 15118 Documents for AC/DC Charging
ISO 15118 Protocol Overview
ISO 15118 Key Features
ISO 15118 Integration
Daimler AG
ISO 15118, Daimler, A. Heinrich, 2016-06-01, Page 2

Why ISO 15118 ?

What does the user want ?

- Availability: As available as a standard car
- Charging: As comfortable as possible

Problem:

- EV needs frequent recharging
- Frequent recharging requires frequent payment/authorization
- Frequent payment/authorization is not comfortable

Example: Cash, Credit Card and RFID payment

Source: S. Voit

Source: A. Heinrich

Daimler AG

ISO 15118, Daimler, A. Heinrich, 2016-06-01, Page 3

Why ISO 15118 ?

What is needed for mass market?

- Grid compatibility: EV must charge with Load Management
- Charge anywhere: Customer can charge anywhere (Roaming: independent from energy provider)

Problem:

- Load Management controlled by grid contradicts customers wish for high availability: „car is charged at any time“
- Comfortable roaming requires standardized payment system but which is the best system?

Example: Standardized RFID

Source: A. Heinrich

Daimler AG

ISO 15118, Daimler, A. Heinrich, 2016-06-01, Page 4

The Solution: Smart Charging

- Use additional information exchange between EV and EVSE

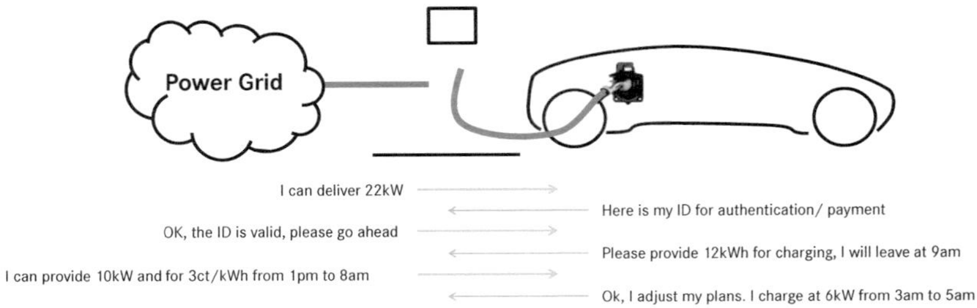

Daimler AG

ISO 15118, Daimler, A. Heinrich, 2016-06-01, Page 5

Problem: What to Standardize?

- Information Exchange requires different interfaces
- Which communication shall be standardized ?

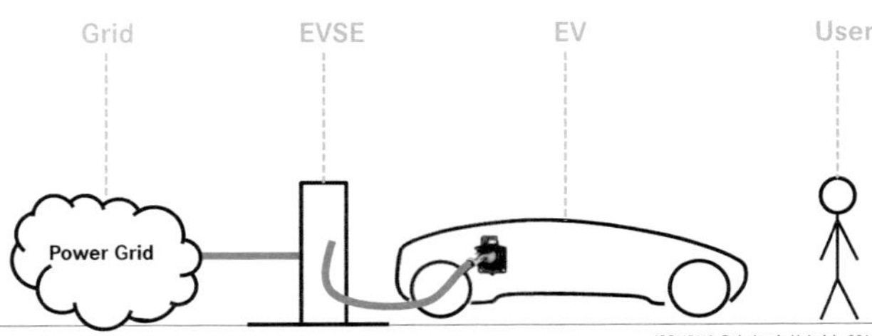

Daimler AG

ISO 15118, Daimler, A. Heinrich, 2016-06-01, Page 6

Problem: What to Standardize?

- OEM: Any protocol, own user interface, own look and feel for user
- Utility/Network Provider/Mobility Operator: any protocol

Example: Please provide 12kWh to charge, I will leave at 9am

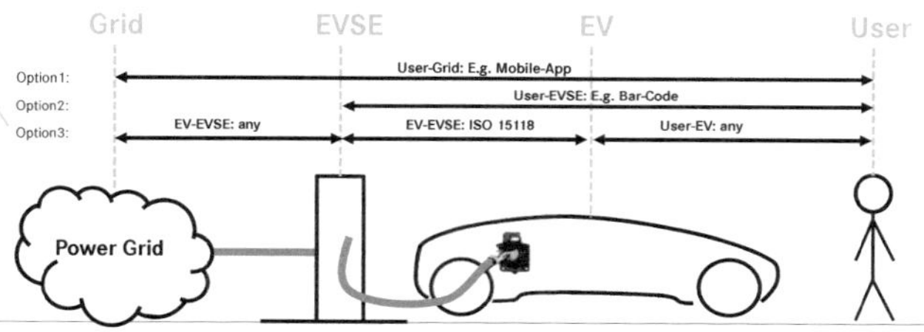

Daimler AG

ISO 15118, Daimler, A. Heinrich, 2016-06-01, Page 7

History of ISO 15118

- **9/2008: Daimler/RWE Project for Smart Charge Communication**
- 3/2009: JWG V2G CI Kick Off
- **9/2009: ISO/IEC new Mode 5 Project registered**
- 6/2010: Project approved
- 4/2011: Committee Draft (CD)
- 5/2012: Draft International Standard (DIS)
- 10/2013: Final Draft International Standard (FDIS)
- **6/2014: International Standard (IS)**
- 6/2015: International Standard to be revised

2009: Smart ED PhII with load management and PnC

Plug&Charge: No Display, No RFID for authentication

Daimler AG

ISO 15118, Daimler, A. Heinrich, ... Page 8

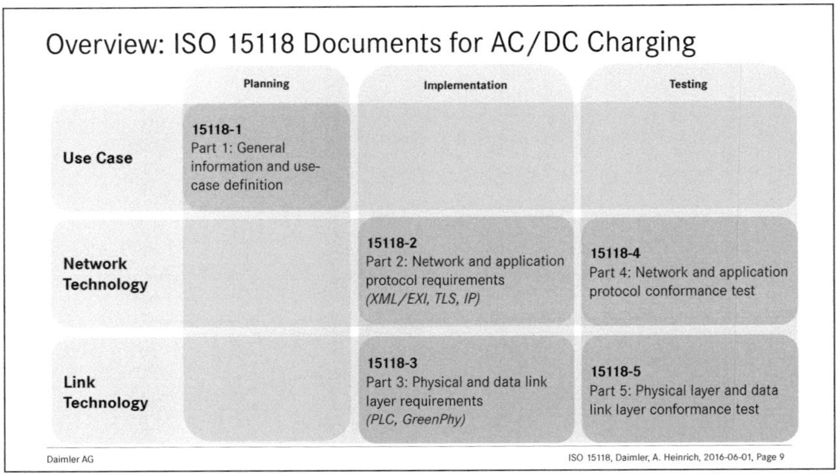
Overview: ISO 15118 Documents for AC/DC Charging
Planning
Implementation
Testing
Use Case
15118-1
Part 1: General information and use-case definition
Network Technology
15118-2
Part 2: Network and application protocol requirements
(XML/EXI, TLS, IP)
15118-4
Part 4: Network and application protocol conformance test
Link Technology
15118-3
Part 3: Physical and data link layer requirements
(PLC, GreenPhy)
15118-5
Part 5: Physical layer and data link layer conformance test
Daimler AG
ISO 15118, Daimler, A. Heinrich, 2016-06-01, Page 9

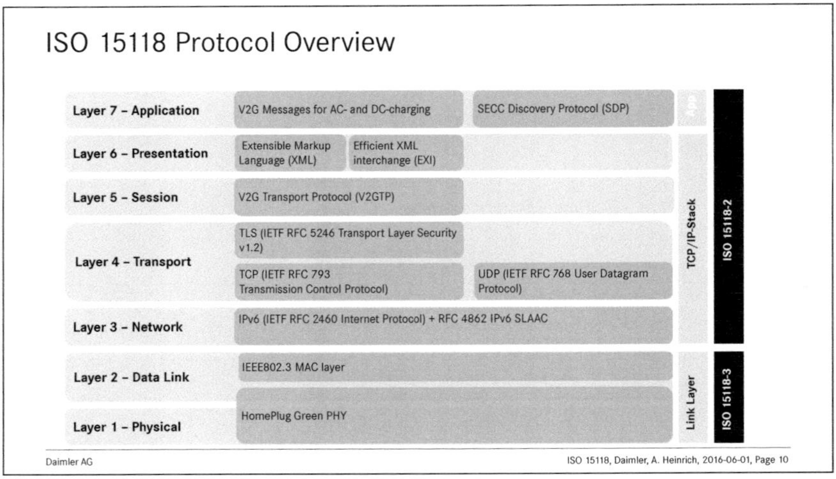
ISO 15118 Protocol Overview
Layer 7 – Application
V2G Messages for AC- and DC-charging
SECC Discovery Protocol (SDP)
Layer 6 – Presentation
Extensible Markup Language (XML)
Efficient XML interchange (EXI)
Layer 5 – Session
V2G Transport Protocol (V2GTP)
Layer 4 – Transport
TLS (IETF RFC 5246 Transport Layer Security v1.2)
TCP (IETF RFC 793 Transmission Control Protocol)
UDP (IETF RFC 768 User Datagram Protocol)
Layer 3 – Network
IPv6 (IETF RFC 2460 Internet Protocol) + RFC 4862 IPv6 SLAAC
Layer 2 – Data Link
IEEE802.3 MAC layer
Layer 1 – Physical
HomePlug Green PHY
TCP/IP-Stack
ISO 15118-2
Link Layer
ISO 15118-3
Daimler AG
ISO 15118, Daimler, A. Heinrich, 2016-06-01, Page 10

ISO 15118 Key Features

	AC	DC
Plug & Charge	• **Convenient**: automatic authentication and billing • **Secure**: payment based on electronic certificates and signatures • **Cost Efficient**: pure SW solution • **Backward compatible**: other payment systems are supported	
Load Management	• **Renewable focused**: supports prediction of consumed power over time • **Fair**: schedule negotiation meets the needs of user AND grid • **User oriented**: controlled charging based on dynamic rates (incentives/tariffs)	
Value Added Services	• **Extensible**: allows for standard internet access and additional Services • **Manageable**: EVSE can control access via service	

Daimler AG

ISO 15118, Daimler, A. Heinrich, 2016-06-01, Page 11

ISO 15118 Integration

- ISO 15118 extends IEC 61851-1 (PWM-Charging)
- ISO 15118 is part of Combined Charging System (CCS)
- Mandatory for DC, optional for AC
- PLC Data is transmitted via PWM pin (no additional pins)
- Supports Type-1 and Type-2 connector

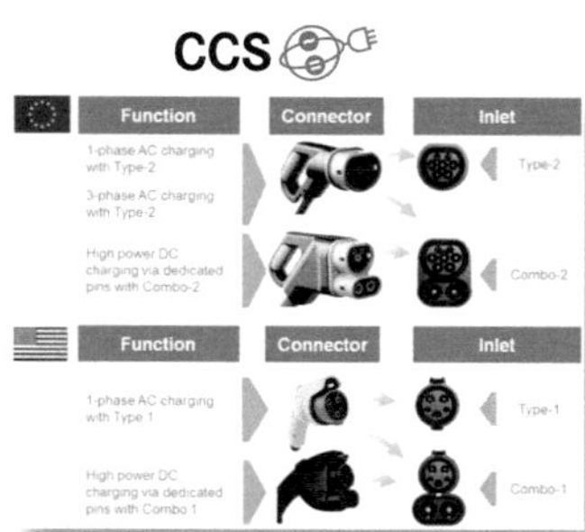

Source: CharIN e.V.

Daimler AG

ISO 15118, Daimler, A. Heinrich, 2016-06-01, Page 12

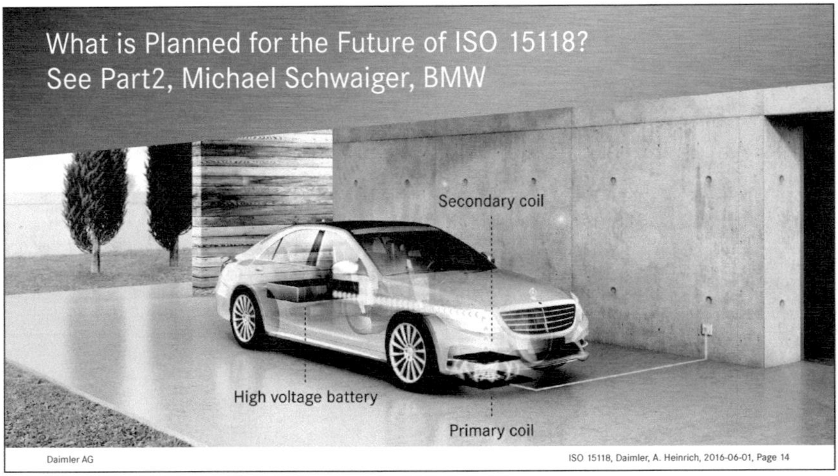
What is Planned for the Future of ISO 15118?
See Part2, Michael Schwaiger, BMW
Secondary coil
High voltage battery
Primary coil
Daimler AG
ISO 15118, Daimler, A. Heinrich, 2016-06-01, Page 14

ISO15118-CHARGING COMMUNICATION

GRID INTEGRATION OF ELECTRIC MOBILITY,
1ST INTERNATIONAL ATZ CONFERENCE
01.06.2016, M. SCHWAIGER

BMW GROUP

ISO15118 TESTIVAL. BACKGROUND.

- ISO15118-1 to 3 are published as International Standards.
- Tests to be done for Interoperability assurance.
- The word Testival is composed of Test and Festival and well known for communication tests of Bluetooth, Zigbee ...
- Companies come together to test communication devices against each other in a round robin manner.
- ISO15118 community held Testivals in Japan, USA and Germany.

Rolls-Royce

ISO15118-Charging communication, BMW Group, 1.6.2016

Page 2

VALIDATION OF ISO15118.
COMMUNICATION TESTIVALS IN 2014-2015.
13-14.11.2014: Testival at ANL in Chicago/USA
International participants, ECUs, vehicles and charging stations tested in round robin
Argonne
NATIONAL LABORATORY
"Argonne National Labs hosts the Interoperability Olympics"
We assumed there would be maybe a couple of vendors bringing small little boards, but it blossomed into this international event.
It was quite surprising that we didn't find any totally un-interoperable pairings
16-17.04.2015: Testival at JARI in Tokyo/Japan
International participants, ECUs tested in round robin
ISO/IEC 15118 V2G CI 互換性試験の
国際標準化活動*
Standardization Activities of ISO/IEC 15118 V2G CI Conformance Tests
Source: eNterop, ANL, Charged Electric Vehicle Magazine
http://chargedevs.com/features/argonne-national-labs-hosts-the-interoperability-olympics/
Source:
http://tech.jsae.or.jp/hanbai/html/readme/kaishi.html
Rolls-Royce
ISO15118-Charging communication, BMW Group, 1.6.2016
Page 3

VALIDATION OF ISO15118.
COMMUNICATION TESTIVALS IN 2015-2016.
12-13.11.2015: Testival at TU Dortmund/Germany
International participants, ECUs, vehicles and charging stations tested in round robin
3rd Testival
ISO/IEC 15118
12th-13th November 2015
Dortmund, Germany
7-8.04.2016: Testival at CSE in Sand Diego/US
International participants, ECUs, vehicles and charging stations tested in round robin
Save the Date!
Sustainable Energy
4th International
ISO/IEC 15118 Interoperability
and Conformance Testing Symposium
April 7–8, 2016
Center for Sustainable Energy
9325 Sky Park Court, Suite 100
San Diego, CA 92123
Invitation with additional details to come shortly
Rolls-Royce
ISO15118-Charging communication, BMW Group, 1.6.2016
Page 4

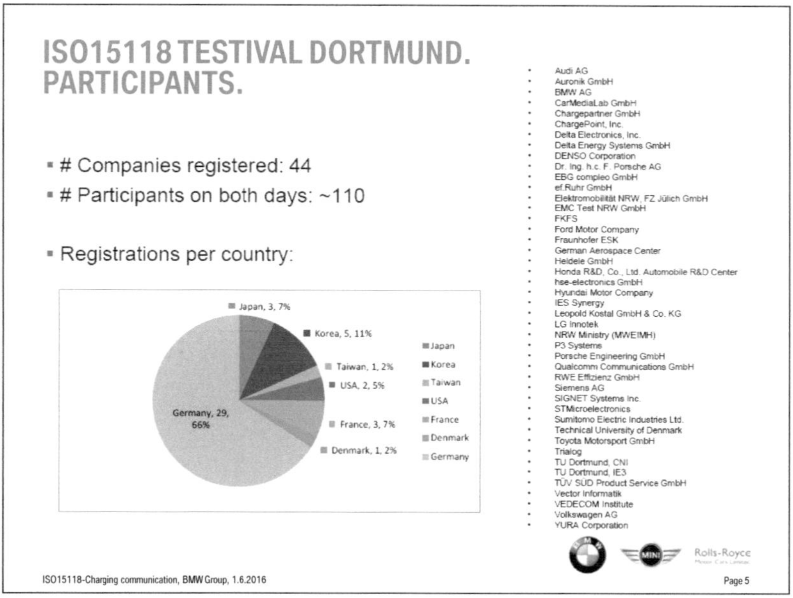
ISO15118 TESTIVAL DORTMUND.
PARTICIPANTS.
▪ # Companies registered: 44
▪ # Participants on both days: ~110
▪ Registrations per country:
Japan, 3, 7%
Korea, 5, 11%
Taiwan, 1, 2%
USA, 2, 5%
France, 3, 7%
Denmark, 1, 2%
Germany, 29, 66%
Japan
Korea
Taiwan
USA
France
Denmark
Germany
• Audi AG
• Auronik GmbH
• BMW AG
• CarMediaLab GmbH
• Chargepartner GmbH
• ChargePoint, Inc.
• Delta Electronics, Inc.
• Delta Energy Systems GmbH
• DENSO Corporation
• Dr. Ing. h.c. F. Porsche AG
• EBG compleo GmbH
• ef.Ruhr GmbH
• Elektromobilität NRW, FZ Jülich GmbH
• EMC Test NRW GmbH
• FKFS
• Ford Motor Company
• Fraunhofer ESK
• German Aerospace Center
• Heldele GmbH
• Honda R&D, Co., Ltd. Automobile R&D Center
• hse-electronics GmbH
• Hyundai Motor Company
• IES Synergy
• Leopold Kostal GmbH & Co. KG
• LG Innotek
• NRW Ministry (MWEIMH)
• P3 Systems
• Porsche Engineering GmbH
• Qualcomm Communications GmbH
• RWE Effizienz GmbH
• Siemens AG
• SIGNET Systems Inc.
• STMicroelectronics
• Sumitomo Electric Industries Ltd.
• Technical University of Denmark
• Toyota Motorsport GmbH
• Trialog
• TU Dortmund, CNI
• TU Dortmund, IE3
• TÜV SÜD Product Service GmbH
• Vector Informatik
• VEDECOM Institute
• Volkswagen AG
• YURA Corporation
Rolls-Royce
ISO15118-Charging communication, BMW Group, 1.6.2016
Page 5

COMBINED CHARGING SYSTEM
CHARGING INTERFACE INITIATIVE ASSOCIATION.
Develop and promote the Combined Charging System (CCS)
Founded in 2015
http://www.charinev.org
CharIN e.V.
JOIN US!
We present the Charging Interface Initiative association (CharIN), which was recently founded to develop and promote the Combined Charging System (CCS).
BMW GROUP
DAIMLER
MENNEKES
PHOENIX CONTACT
Rolls-Royce
ISO15118-Charging communication, BMW Group, 1.6.2016
Page 6

COMBINED CHARGING SYSTEM
DC CCS CHARGING STATION ROLLOUT EUROPE.

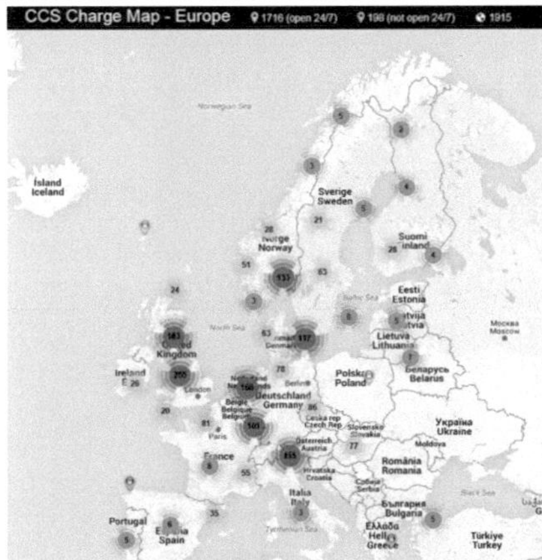

Source: http://ccs-map.eu

>1000 DC CCS charger installed in Europe

ISO15118-Charging communication, BMW Group, 1.6.2016

Page 7

ISO15118.
FUTURE USE-CASES.

Vehicle to Grid communication

1 WPT-charging control

- **WPT (Wireless Power Transfer)** for convenient customer experience
- **One wireless communication channel** for fine positioning, pairing and charge control
- **Authentication of off-board charging equipment** via same methods used for AC and DC charging

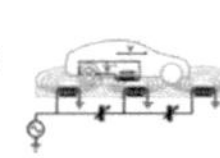

2 Reverse Power Flow

- **Reverse Power flow** for smart grid support
- **Charge-control** via control & status commands
- **Re-usage of communication technology** for AC & DC & WPT charging (i.e. single interface)

3 Automated Connecting Device

- **Support for electric Busses** for public transport
- **Control of Pantograph** for connect/disconnect
- **Short-time, high Power DC charging** at public bus stations

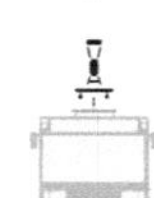

Source: www.designnews.com; www.m2global.com

ISO15118-Charging communication, BMW Group, 1.6.2016

Page 8

ISO15118 - WIRELESS POWER TRANSFER. CONVENIENT CUSTOMER EXPERIENCE.

1 Wireless communication with inductive charging

- Wireless communication (WLAN) setup during Ground Pad approach
- Fine Positioning with auxiliary signal for correct adjustment of Vehicle- and Ground Pad
- Wireless Power Transfer with >90% efficiency

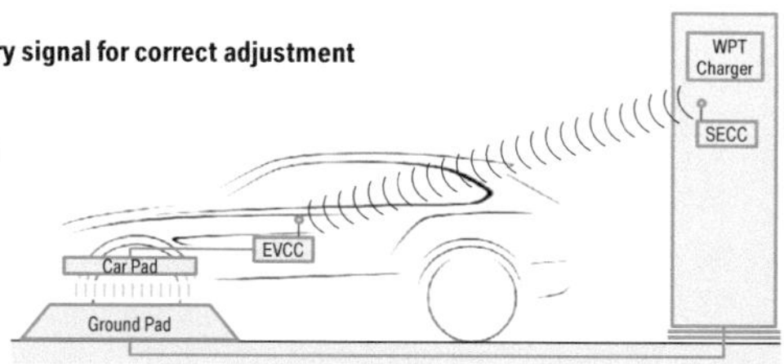

Source: www.autogefuehl.de

Rolls-Royce

ISO15118-Charging communication, BMW Group, 1.6.2016

Page 9

ISO15118 - REVERSE POWER TRANSFER. ADDITIONAL SUPPORT FOR GRID BALANCING.

2 Reverse Power Transfer

- AC, DC or WPT Reverse Power Transfer
- Vehicle as Distributed Energy Resource
- Vehicle to Home (V2H) for emergency power supply
- Vehicle to Grid (V2G) for grid balancing and peak shaving
- Vehicle to Vehicle (V2V) for roadside charging (e.g. ADAC)

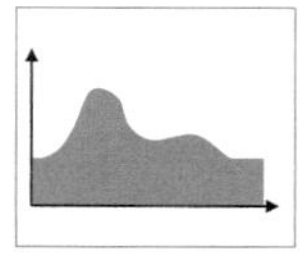

Rolls-Royce

ISO15118-Charging communication, BMW Group, 1.6.2016

Page 10

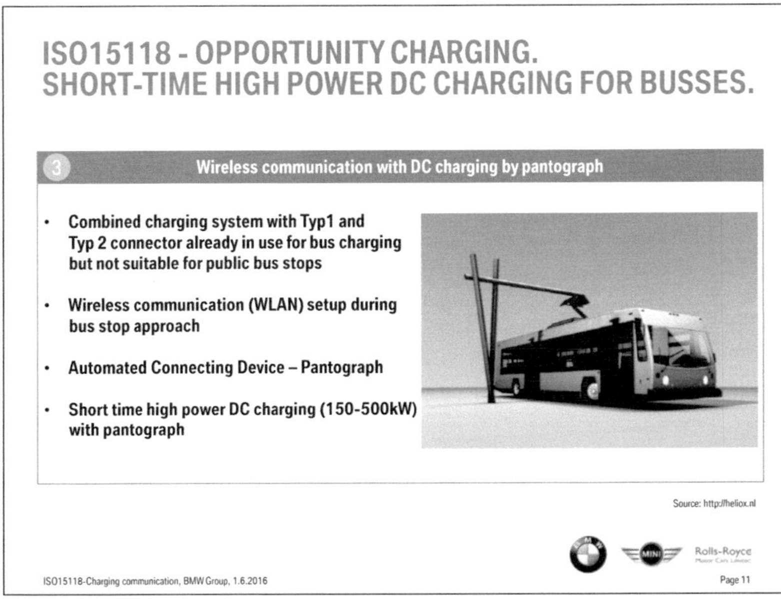
ISO15118 - OPPORTUNITY CHARGING.
SHORT-TIME HIGH POWER DC CHARGING FOR BUSSES.
3
Wireless communication with DC charging by pantograph
• Combined charging system with Typ1 and Typ 2 connector already in use for bus charging but not suitable for public bus stops
• Wireless communication (WLAN) setup during bus stop approach
• Automated Connecting Device – Pantograph
• Short time high power DC charging (150-500kW) with pantograph
Source: http://heliox.nl
ISO15118-Charging communication, BMW Group, 1.6.2016
Page 11

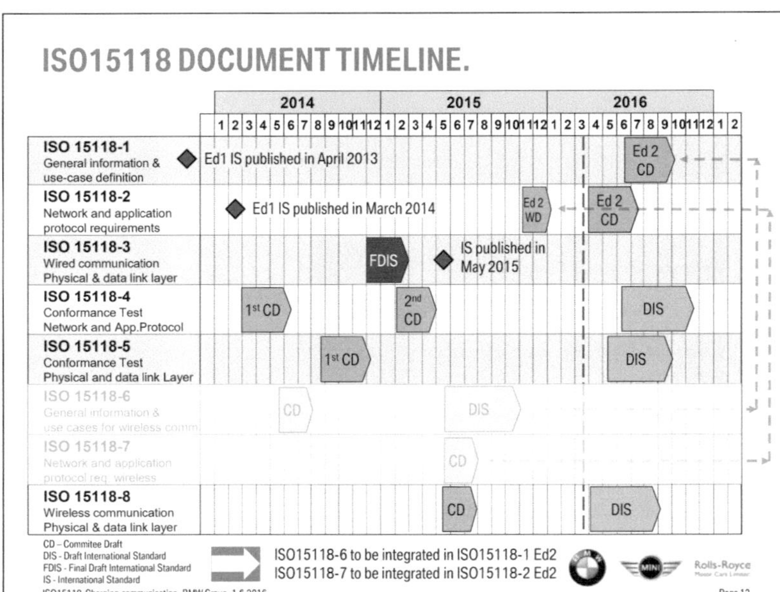
ISO15118 DOCUMENT TIMELINE.
2014
2015
2016
ISO 15118-1
General information & use-case definition
Ed1 IS published in April 2013
Ed 2 CD
ISO 15118-2
Network and application protocol requirements
Ed1 IS published in March 2014
Ed 2 WD
Ed 2 CD
ISO 15118-3
Wired communication Physical & data link layer
FDIS
IS published in May 2015
ISO 15118-4
Conformance Test Network and App.Protocol
1st CD
2nd CD
DIS
ISO 15118-5
Conformance Test Physical and data link Layer
1st CD
DIS
ISO 15118-6
General information & use cases for wireless comm.
CD
DIS
ISO 15118-7
Network and application protocol req. wireless
CD
ISO 15118-8
Wireless communication Physical & data link layer
CD
DIS
CD – Commitee Draft
DIS - Draft International Standard
FDIS - Final Draft International Standard
IS - International Standard
ISO15118-6 to be integrated in ISO15118-1 Ed2
ISO15118-7 to be integrated in ISO15118-2 Ed2
ISO15118-Charging communication, BMW Group, 1.6.2016
Page 12

THANK YOU.
Rolls-Royce
ISO15118-Charging communication, BMW Group, 1.6.2016
Page 13

E-mobility – a challenge for IT as well!

Dr. Ralf Walther, Rolf Norrenbrock, Heike Lüschen, Malte Neuendorf

Abstract

The market launch of electric mobility (eMobility) coincides with the megatrends energy transition and digitization. Therefore the launch takes place in a dynamically changing energy market and increasing automation of processes and digital networking systems. This transformation occurs on timescale of the energy industry very fast. But also for the information technology, the pace of innovation is very high. As a consequence of this IT infrastructure and applications should be designed modular and scalable, and future requirements have to be already anticipated. A further challenge is the need of cost efficiency, which arises for example in the conflict between IT security / safety and the resulting costs. The German Bundestag has adopted the draft law to increase the security of IT systems. It is expected that the strict requirements, which are already known in the field of energy grid control, has to be extended to the entire energy operational infrastructure. Concerning the possible development paths, economically viable intermediate milestones as well as the resulting scaling costs are important aspects. In order to be successful at the impending rollout and in order to be able to concentrate on the core business, the main decisions for a sustainable infrastructure should be made today.

eMobility: Driven by many Megatrends!

Megatrends are long-term and comprehensive change processes, effective powerful influences that shape the markets of the future. They differ from other trends by their long time horizon, by their reach and their potency. Figure 1 shows a selection of the mostly named Megatrends – lots of them (the blue ones) are strongly correlated with electric mobility.

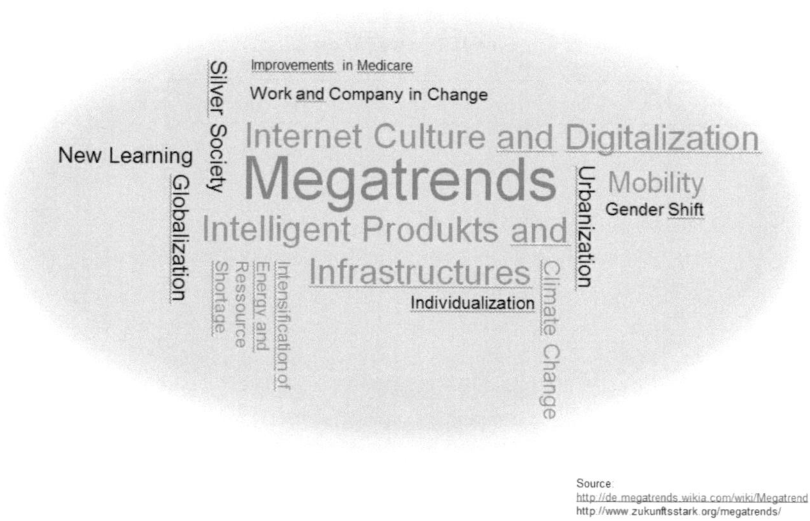

Figure 1: Mega Trends

A deeper look at these trends give us an impression of the cross-linking to other fields of life. The car of the future is not only driving on electric power. It will also communicate e.g. with other cars, transport infrastructure and the Internet. Even the car manufacturers and energy service providers will be in permanent communication with the car and offer in addition to current services a variety of other new services. This market of services is that attractive that the conventional market roles (see also figure 2) are becoming increasingly blurred. Car manufacturers are no longer just making cars, energy providers are no longer just selling energy and internet companies are no longer just share information. And all of them try to get access to the personal data – as basis for further services.

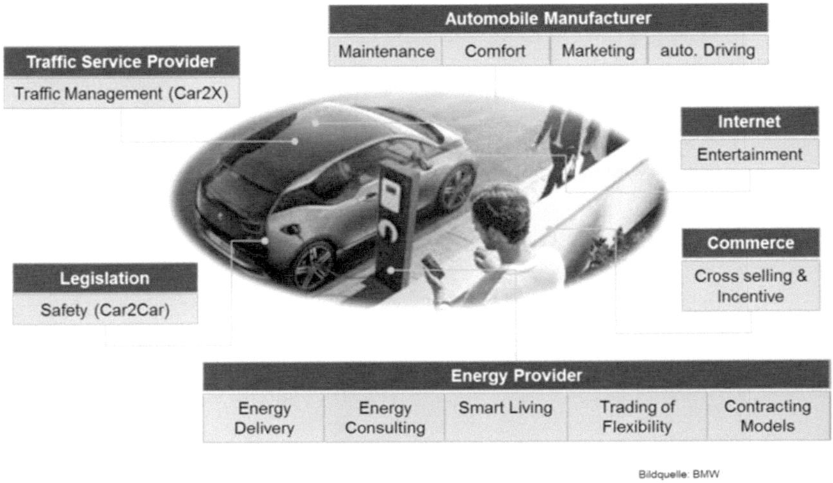

Figure 2: Overview of rules and services. In future, the conventional market roles will increasingly blurred.

In the following part of this article I will focus on the role and services of the energy provider. From its traditional core business the following services are resulting: energy supply, energy consulting, Contracting, Smart Living (energy, IT and telecommunication) and trading of flexibility. In this connection the energy supplier benefits of its customer focus, efficient energy market processes and last but not least the energy market access.

Looking at the different services you recognize a different time to market. For example, the services "Energy Delivery" for electric mobility are already implemented. Business models for contracting and Smart Living are short before market entry. Technically sophisticated services and business models that needs a larger market penetration are more likely in the medium-term future.

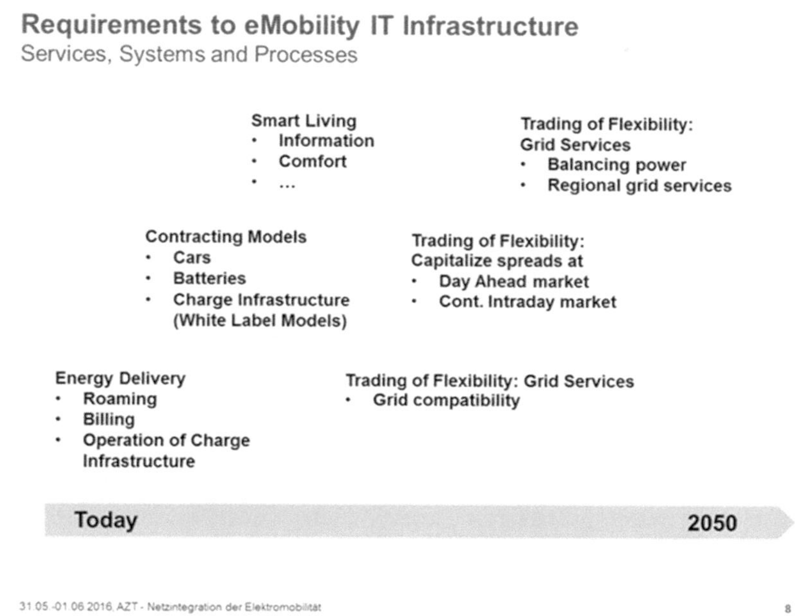

Figure 3: Time to market for different services.

However, all these different services should be already thought in the design of the IT infrastructure in order to be able to react quickly and flexibly to the market and not to get into an IT technical impasse, which would result in increased costs in the future.

Requirements of IT Infrastructure

These different services in figure 3 have different impact on IT requirements. Simplified it can be said that a conflict between the different requirements and system cost results, at which the hardest requirement dictates the requirement for the entire system. Having in mind the different time to market of the certain services, an economically viable development path becomes a challenge for IT infrastructure. In this consideration the grid services are standing out with very strict requirements. From today's perspective and with the current legal conditions an economic grid service can be hardly realized without smart meter.

Service	Systems / Processes	Security	Operational availability	Performance	Costs
Energy Delivery	• Roaming • Billing • Operation of Charge IS	+	~95%	+	+
Contracting Models	• Cars • Batteries • Charge Infrastructure (White Label Models)	+	~95%	+	+
Energy Consulting	• Process Analysis • Internal Process Opt.	+	Offline – 95%	++	0
Smart Living (Vernetzung und smarte Produkte)	• Information • Comfort • ...	+	95%	+	+
Trading of Flexibility	Capitalize spreads at • Day Ahead market • Cont. Intraday market	+	>99%	+	+
	Grid Services • Grid compatibility • Balancing power • Regional grid services	++		++	++

Figure 4: Technical requirements of different services.

Among the non-functional requirements the attack security (Security), operational safety (Safety) and data protection (privacy protection) are of particular note. In the recent past there have been a number of incidents in which an hacker managed to get the remote control over the IT system of a vehicle or IT system. This is not a specific problem of electric mobility but the increasing networking of automobiles as such. In the field of power supply systems, especially the novel "Black Out" has very figuratively pointed to the vulnerability of modern society to attacks on energy infrastructure. The legislature responded to the increased protection required by more stringent legal requirements, such as the IT security law. Even data protection is not a special challenge of electronic mobility but also a result of the increasing digitalization. Again, the society is already sensitized. The legislature responded again with more stringent requirements.

Another non-functional requirement for the IT infrastructure arises from the cost pressure and the strong internationalization of the automotive industry. In order to avoid high interface costs, the definition of international standards are of enormous importance. Here already established bodies and institutions such as IEC and VDE are engaged. But in future it will also depend on being able to keep pace with the fast changing technical and market-related requirements. Again, it requires a great amount of anticipation ability.

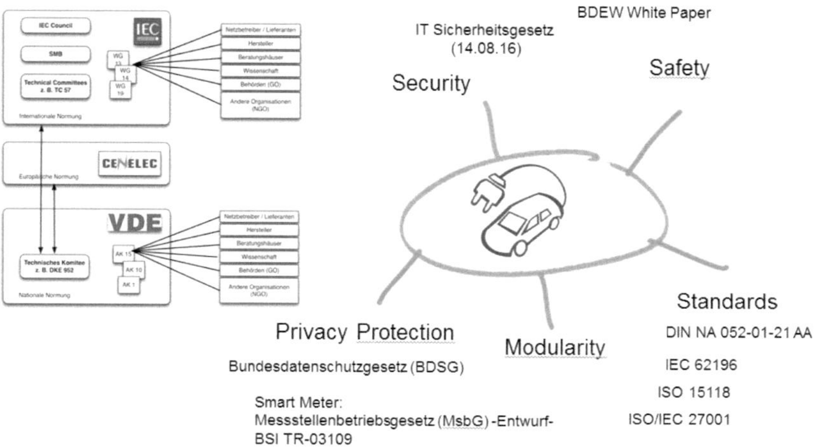

Figure 5: Non-functional requirements: Standards.

Moreover, there is a certain complexity arising from the diversity of the relevant processes and technical depth.

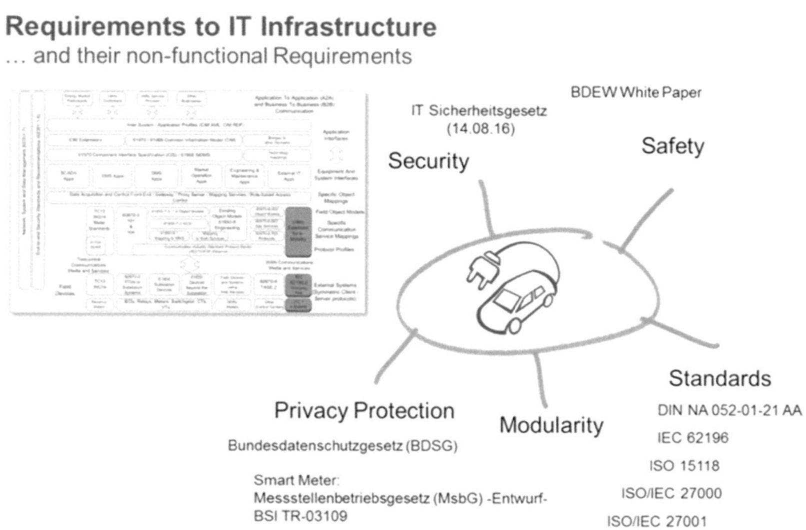

Figure 6: Non-functional requirements: technical diversity.

Functional Placement / System Overview

All these requirements lead to a modular, flexible system architecture. In addition to the technical structure, e.g. safety standards of the data center, hardware performance, safety zone concepts, interface concepts, etc., the so-called functional placement is very important. A rough example is shown in figure 7. Here the different layers and functional groups are implied, which should be considered when building an IT infrastructure. These functional groups will of course vary depending on the orientation and the core business of the energy services provider. A corresponding function Map results after analysis of the relevant services and relevant business processes.

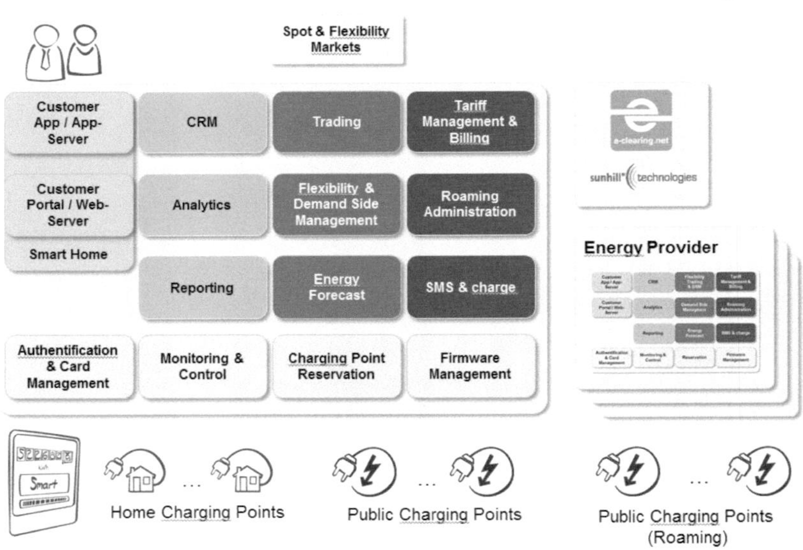

Figure 7: Non-functional requirements: technical diversity.

SYSTEM BEHAVIOR

The grid friendly integration of shiftable loads – the approaches from the EnBW Pilot Project "Flexible Power-to-Heat" also suitable for electric vehicles

Holger Wiechmann

1 Introduction

The German energy transition ("Energiewende") is one of the central tasks, to which politics and society have committed themselves. As part of the energy transition, the fluctuating renewable energies (RE) such as wind and solar in particular will be massively expanded and there will generally be a stronger electrification in almost all economic sectors – including the transportation sector – through the substitution of fossil energy. To accommodate high amounts of fluctuating RE within the energy system, innovative approaches to improve the balance of generation and demand must be developed in particular. Load management and the issue of utilising flexibility will play a key role. The pilot project "Flexible power-to-heat" – a joint project between EnBW and Netze BW – addresses these aspects and could also be an example for a grid friendly charging of electric vehicles.

The issue of flexibility within power demand is currently being discussed intensively within the energy sector. In the context of the upcoming EnWG (energy industry law) amendment regarding a new energy market design incorporating the challenges of the energy transition, the issue of flexibility has become one of the central points [BMWi 1]. The specific implementation approaches dealing with how the flexibilities can be best utilized are of crucial importance. Demand-side flexibility potential is generally available to a very significant extent; it just wasn't harnessed, because until now it has not been necessary.

With the pilot project "Flexible power-to-heat", EnBW has developed and tested a coherent concept, suitable for mass marketization, to increase flexibility potentials. In particular, the focus is on efficient interaction between regulated and market areas. The principles of the grid traffic light classification system [BDEW 1], which is accepted industry-wide and was co-developed by EnBW, form the starting point for development. Within the trial flexible, discrimination-free, market-driven and network-beneficial load management was implemented using the concrete example of existing electric heating devices. It could be shown that this kind of load management approach – similar to a virtual power plant – is also possible within this highly fragmented mass-market customer segment. At the same time, the implemented flexibility approaches can also be used in principle for many more applications as the charging of electric vehicles and batteries or even (small) CHP or other generation units.

The aim of the pilot project is to allow as much flexibility and potential for load transfer by any supplier, whilst avoiding local overloads in the distribution grid. Implementation of the concept with respect to network-usefulness takes place in the form of a new quota system (QS). These represent the load limit of the respective network section, within which potential suppliers can operate freely and shift energy consumption. The distribution grid operator non-discriminatorily provides all market participants with these QS factors based on projections and calculations. The QS approach therefore offers degrees of freedom to the market, which allows market-driven actions, for example, optimisation of procurement, while safeguarding the convenience of the grid.

2 Utilisation of flexibility – status quo and potential

2.1 Initial situation: Emerging networked worlds

The conversion of the production structure due to the energy transition in Germany is equivalent to a paradigm shift: Starting from several hundred mostly thermal power plants, as well as a limited number of hydroelectric power stations, which historically were (and still are) owned by only a few utilities, decentralized and small-scale production units are being operated increasingly by private companies like domestic and industrial customers. There are now well over a million "power plant operators", who own and operate a majority of PV and wind energy plants. In recent years existing utilities have begun to invest strongly in RE systems. Apart from offshore wind energy, which is viewed as central production technology, resources are being invested in decentralized distribution systems.

A quick look at the development path pursued by the politics for fluctuating RE also shows to what extent new generation capacity will be expanded and what capacities are required for a (large scale) transition to RE supply. Currently, the installed capacity of wind and PV systems totals about 90 GW [BMWi 2]. To achieve the political targets of at least 80% RE in the energy network by 2050, far higher installed generation units of fluctuating REs of up to 250 GW are necessary [fig 1]. These additional capacities will be mainly wind and PV. The actual shortfall is about 140 GW with a share of 50 % of Wind and PV each.

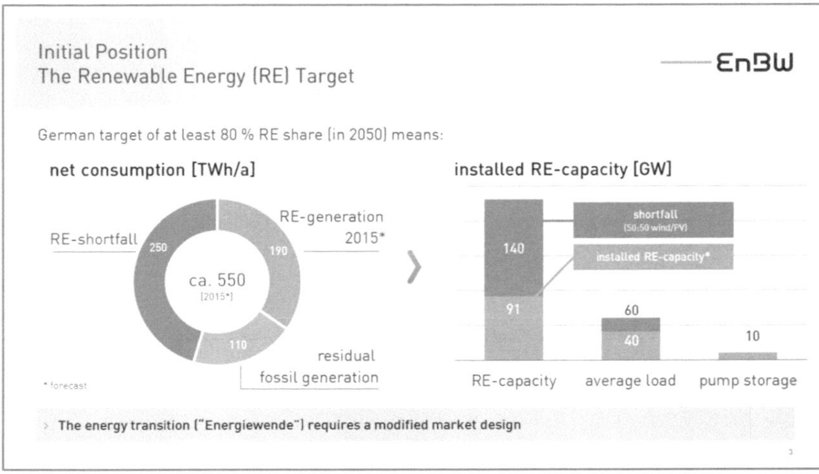

Figure 1: The Renewable Energy Target

To better classify these numbers, the "usual" figures for energy power demand in Germany provide a good indicator. Generally this varies daily in terms of bandwidth between 40 and 60 GW, with certain differences between workdays (high load) and weekends (low load). During (winter) peak times, the maximum demand load has remained slightly above 80 GW for years now and the usual minimum load occurs on Whit Monday with about 30 GW [dena 1].

In addition, there are cross-border grid capacities of 21 GW, which can be used for export or import, 11.6 GW pumping capacity in the German control area [BNetzA 1] and a must-run generation from thermal power plants and CHP plants, which currently lies between about 15 GW (summer) and 25 GW (winter, through heat-guided CHP amongst others) . The first two increase the potential demand – even if only on an hourly basis – the latter reduce the demand balance by the corresponding must-run generation.

Conclusion: The numbers for the installed generation capacity show that the existing energy system will soon be at its limits, if further measures such as, for example, greater use of flexibility are not implemented. Intelligent flexibility management also has the potential to take full advantage of existing network capacity or to help avoid bottleneck situations. This smart control of demand and decentralized generation/storage comes under the heading of 'virtual power plant'.

2.2 Current and future flexibility potential

The basic flexibility potential on the demand side in Germany is significant. Therefore, [VDE 1] and [Schäfers 1] both assume an industrial load transfer potential, which in theory lay at around 25 GW in 2010 and will double by 2030 to approximately 50.5 GW. Our own studies support these statements. Distribution ability per unit time or flexibility is about one hour up to several days in the case of a targeted product shift of non-time critical goods.

In addition to the industrial potentials described, flexibilities also exist at the level of the domestic consumer. In particular all thermal power applications, which store heat temporarily or where heat demand can be shifted in time, are theoretically suitable for flexible operation. In contrast to direct heaters and instantaneous water heaters, many heating and hot water systems have storage units, and thus could be loaded flexibly. In particular this applies to storage heaters, which have so far been loaded mainly at night and which supply the stored heat during the day. These so-called §14a facilities – which include heat pumps and in future electric vehicles – are also called interruptible loads (IL), which may only be activated at defined times each day. Within the relevant retail segment, the installed storage heater and heat pump capacity in Germany totals around 30 GW [Walton 1]. Distribution ability per unit time or flexibility is about several hours up to 48 hrs within the transitional period varies greatly with season.

The Customers Position
Huge investments in future load shift potentials

EnBW

Three examples of local flexibility and load management potentials:

- (Small scale) stationary battery systems
  - Assumption: 50 % of the residential buildings (10 m) with battery system (aver. capacity of 10 kW)
  - Additional capacity of 100 GW
- (Small scale) moveable battery storage (e-mobility)
  - Assumption: 50 % as e-vehicles (25 m) with aver. charge capacity of 20 kW
  - Additional capacity of 500 GW
- Heat storage (hybrid heating)
  - Heating rod with a backup heating system based on gas, oil, heat pumps, etc.
  - Assumption: 50 % of the residential buildings (10 m) with a heating rod (average capacity of 10 kW)
  - Additional capacity of 100 GW

Conclusion: Small scale storage will be an important jigsaw piece of the energy transition!

Figure 2: Huge investments in future load shift potentials

In the medium to long-term, the described flexibility potential in Germany will increase over the next few years significantly as hybrid heating systems are increasingly used. Compared with conventional, generally oil or gas-fired boilers, these systems have an additional electrically-powered heating element. Battery storage and of course electric vehicles will also increase this potential through their market penetration. Long term, an additional potential up to 700 GW is possible (fig 2). Also in industry, extra flexibility potential can be tapped through additional energy management systems. With the additional flexibility potential, distribution ability per unit time or flexibility should be about several hours up to 48 hrs.

Conclusion: Flexibilities already exist in the current energy system, however they are still neither needed nor being used. In the long term, however, there is a need to raise this potential due to the expansion of fluctuating energy generators. New applications such as electric vehicles and battery storage in the future will further increase the distribution potential extremely. In particular it becomes apparent that these will have to be operated by intelligent systems due to the greater installed power to prevent system errors.

3 Requirements and resulting solutions approach

3.1 Requirements of the stakeholders

To use flexibilities on the consumer side in the best way, two requirements must be met. First, flexibilities should make a contribution towards integrating increasingly decentralized power generation as effectively as possible into the current energy system. On the other hand, it must be ensured that the use of flexibilities does not lead to new problems due to high power load. So the DSO's are one of the main stakeholders.

The previous system of integration of electric heating devices (EHD) in the distribution network aimed to exploit existing electricity networks as effectively as possible and prevent grid bottlenecks. In particular, the load groups instrument was used to place EHD consumption within off-peak hours (e.g. at night). The reliable and predictable behavior of the ILs is the basic prerequisite for the adherence of network load limits and thus the prevention of alternative network expansion. Even with increasing flexibility of consumers, compliance with the requirements of the responsible distribution network operator regarding maximum concurrent loads must be ensured.

In return for their network-relevant behavior, customers receive a separate network charge for the power consumption of their system. This principle must be maintained in order to ensure integration of large, controllable consumers that benefit the network in the future.

The point of view of sales companies is fairly simple to sum up: "We want (to be able to) offer our customers attractive products and tariffs and thereby the option to make a contribution to the energy transition". In essence, it comes down to being able to connect production and consumption with each other flexibly.

In the mass-market segment with its static load profile procedure, but with a large existing and future increasing flexibility potential, this is not (economically) viable to implement due to insufficient regulatory guidelines. To flexibly implement energy-logistics processes, new accounting rules, market processes and a newly designed §14a are required in particular. This requires the integration of the needs of the network operator or the best possible use of available network capacity.

From the point of view of sales and marketing, it comes down to establishing a regulatory framework for flexible products. The EnBW is willing to offer flexible energy products.

3.2 Resulting solution approach – the quota system

The aim of the approach is to avoid local overloads in the distribution network and grid bottlenecks, e.g. through simultaneous loading of all ILs in a network, whilst allowing as much flexibility and potential for load transfer by any supplier. To do this, previous load groups are replaced by a quote system (QS).

The grid operator sets quotas for the entire subordinate area of a transformer of high/ medium voltage. These stipulate what percentage of the nominal power of the locally installed, controllable consumers may take electricity from the grid at the same time. This percentage applies without discrimination to all suppliers being active in their respective network. The restrictions must also be determined individually for each network level, i.e. different quotas at the level of a medium voltage circuit, a local network station or a low voltage circuit.

Unlike existing load groups, these no longer apply to a single device, but for an entire system collective within a network section. Network-side boundary conditions – i.e. maximum concurrency of activation – are still respected and, at the same time, opportunities for suppliers are created to transfer loading of IL.

Example: The network operator sets a quota of 0.5 for the area of a transformer of high/ medium voltage. Then all suppliers in the area of this transformer may access 50% of the IL output under contract with them. This may mean that in the area of one power station all IL are switched on, whilst in the area of another power station none operates. To ensure that there are no local overloads, the operators set additional conditions if necessary. So, for example, this might mean that for a local network station only 80% of interruptible power installed in the supply area of the local network station may be used at the same time, for a low-voltage circuit only 60%, for another over 100% of the interruptible load connected to this circuit.

To determine the quotas for a network area, different approaches can be applied:

Convert existing load groups to quotas based on today's enable / red periods,

Performance of network simulations to quantify the theoretically available free capacity at each network level

Consideration of real measured data (e.g. at substations, smart meter measurement) to quantify real, available free network capacity at each network level

Quotas should be formulated on a temporary basis to take account of fluctuations in consumption or production (seasonally as well as within one day). Local production may also be considered in an additional step which may lead to higher quotas, especially at lunchtime.

While the first approach resolves the problem of load group mapping, which was previously per device, the load transfer potential still remains limited. However, this approach offers the advantage of fast implementation.

Performing network simulations is expensive, but results in a clear increase in load transfer potential, even in compliance with existing network planning premises. Network models and profiles were created on the basis of the integrated data basis. The subsequent determination of quotas is performed in an iterative process. Starting from full activation of all IL, network overloads are recorded and activation reduced successively, until there are no more overloads.

Within the pilot project, quotas with a resolution of ¼-hour were formulated for every transformer at the levels of the low-voltage strand, local network stations, as well as the medium voltage strand. The necessary simulations were performed as part of a research partnership with the Institut für Hochspannungstechnik of the university RWTH Aachen.

Figure 3 shows the comparison between the behavior of real IL-devices in their previous clearance and breaking periods (orange) for an exemplary local network, as well as the simulated possible clearance quantities within the context of establishing quotas (blue). In total, the electric heating devices with a total capacity of 320 kW are installed in the local network under consideration; the local network transformer is however only measured up to a maximum load of 250 kVA. Thus, the previous system of enable and red periods already has the effect of limiting the concurrency on the consumer side.

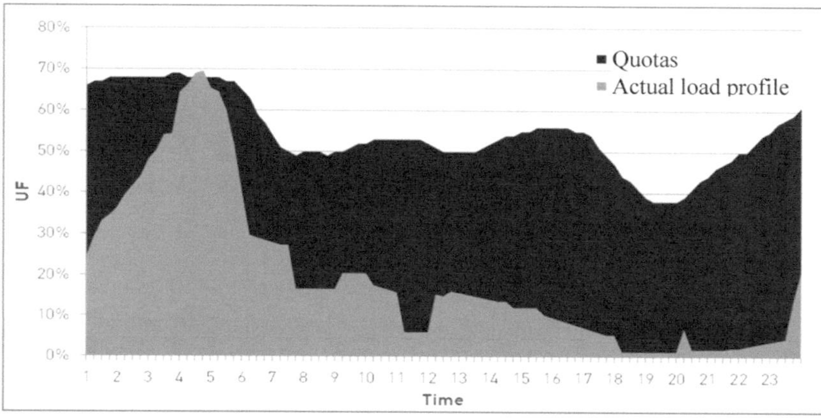

Figure 3: Exemplary application of quotas, compare with previous behavior

It is clear that in the previous system a great deal of flexibility is not used in the distribution network (see blue area). By transferring power consumption in these periods, procurement benefits are increased within current network restrictions. Network expansion can also be partially avoided, if flexibilities on the consumer side are used intelligently or so as to benefit the network.

This quota system is our suggestion for a new § 14a. The so called § 14a law gives the basic for special grid fees for interruptible application like electric heat, electric vehicles or other interruptible devices.

The measured values from the electrical substation and stations are also being used to develop an emergency stop function. This will serve as a back-up level, in case overloads occur in specific areas of the grid. At each low voltage outlet and the transformer in the measured local network stations, maximum values for the current capacity are set, which total 90% of the relevant backup size. If one of the real time measurements exceeds this limit, an emergency-off signal is sent to a central server automatically, resulting in the shut-down for at least an hour of all electric heating devices connected to the affected operating to reduce the load in the network section. A similar emergency off is implemented in the Netze BW control center, which is connected to all systems in the pilot project and then comes into play in event of a surge in the substation transformer. Several function tests of the emergency stop function were completed successfully; use during operation has so far not been necessary, because the use of resources at all monitored network levels has been under the limit. This emergency stop function is also a part of the grid traffic light classification system (red colour).

4 Overview over the pilot project

With the selection of the network areas (fig. 4), the focus lay especially on the acquisition and refitting of customer equipment. 20 customers were accepted into the pilot, during the first heating period in 2013/14. A year later, this was increased to 150 customers.

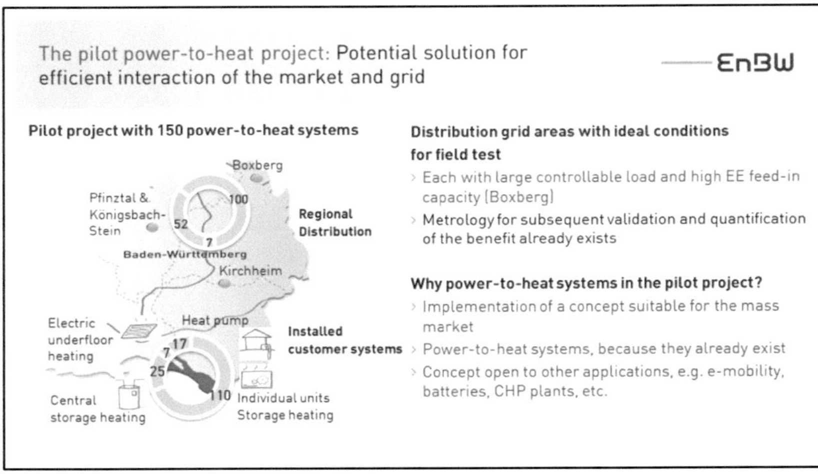

Figure 4: The pilot power-to-heat project

All participating customers have received a smart meter, a control box, a link to the customer's own communication equipment and – for selected customers – several temperature sensors. To obtain targeted direct responses to control signals, all customers were upgraded in case their electric heating device had a so called forward control. To respond to possible failures of communication, emergency schedules were installed on the control boxes to further ensure adequate supply to customers even in event of a communication failure. This included installing the charge schedule of the previous management of the distribution network operator.

One of the lessons learned from the pilot project is that installing modern measuring technology – this applies especially to the pending smart meter rollout – is sometimes connected with unexpectedly high overheads. This is due to the complex situations found in and around the meter cabinets

The mathematical algorithms for intelligent control of customer equipment, as well as determining demand for heating energy represent the core elements of the sales-side. The idea is based on a concept for a new systems approach for market and production-oriented regulation of demand via centralized load management [Wiechmann 1]. The whole system approach has now been developed for practical application within the pilot project, and was tested on real systems.

In a first step, the daily energy requirement of each electric heating device is predicted. This also includes determination of the individual flexibility parameters, i.e. at what time and which amount of energy do the electric heating devices need, such that, for example, individual comfort criteria's (minimal and maximal temperature levels) of customers are met.

The latest weather forecasts are applied to predict each customer's heating energy demand the day before delivery. As mentioned above, this forecast is a key determinant to predict each customer's flexibility. Combining the network's restrictions with its quotas and the customer's potential of flexible operation capacity, we yield possible operating schedules.

The consumer's optimal operation schedule is identified in regard of market price expectations on the energy stock exchange constrained by thermal and network restrictions (based on the UF). This approach is applicable and appropriate, for both the hourly day ahead auction and the quarter-hourly intraday auction. Potentials can be obtained in both markets compared with the status quo.

One of the key challenges in this context is the division of the energy consumption of customers into shiftable and non-shiftable units. This is due to the fact that in addition to the actual electricity consumption for heating, "ancillary consumption" may occur. As the greatest shift potential occurs for heating energy requirements, the pilot project has focused on this.

The advantage of implementing hot water preparation or additional, future applications, e.g. electric vehicles and battery storage, lies in the year-round use and an extension of the flexibility potential. These applications will be implemented in future steps.

The first step of planning and purchase the sum of the individual daily energy amount, results in the total daily energy amount. Another algorithm module determines the best schedule from a combination of flexibility parameters, price expectations (e.g. stock market prices or balancing power) and potential network restrictions on the part of the DSO (quotas).

The schedule produced in this way, is then "translated" again into customer-specific schedules and sent to the individual electric heating devices. During the day, a review of the daily energy quantities actually consumed is performed regularly to ensure as much as possible that only the purchased quantities of energy are consumed or to enable short-term adjustments to be made by intraday trades. If a review finds an unacceptably high deviation, then it can be re-adjusted over the 24 hr period.

Summarized results

The results of the pilot project can be summarized as follows:

Over 100 customer installations were optimised regularly during the pilot project. In fact, there have been over 5,000 accumulated customer optimisations since December 2014 with targeted load shifts, which have demonstrated the actual functionality of the system. Evaluations have shown that the given network restrictions can be reliably maintained by the quotas.

From a sales perspective, focus must be placed on the "product potential" of the approach. Thanks to new tariff products, the rapidly increasing amounts of electricity from fluctuating renewable energy sources as well as additional demand services such as hybrid heating systems, battery storage, or E-mobility can be integrated efficiently in the energy system. These innovative tariff products could offer advantages when compared to the previous tariff offers for flexible customers.

From the perspective of the distribution network operator, Netze BW, it was demonstrated that additional flexibility potential can be raised using the QS approach without overloading the existing grids. Although the risk of potential congestion might increase, it can be managed with appropriate measuring technology (smart grid). The QS approach could be a model for a more advanced §14a EnWG for interruptible loads.

From a customer perspective, it was demonstrated that the feeling of comfort regarding heat supply was still reliably operated, or, with frequent daily loads, even improved.

In addition, it is most likely that more electric vehicles, battery storage, hybrid heating systems and automation systems will enter the market in the future. Therefore, the introduction of mass market load management systems is inevitable from an energy-logistics as well as network capacity point of view to ensure system stability. This means that the focus is on these great flexibility potentials for commercial and household customers.

Especially smart home solutions and energy management systems, with their equivalent or similar control algorithms, represent a danger for system stability due to their swarm-like behaviour (higher concurrency than previously). On the other hand, these systems are particularly suitable thanks to their control potential, for integration in system-side central load and flexibility management. With the solutions approaches developed and tested within the pilot, EnBW is now basically in a position to develop load management suitable for mass market application.

But to be able to implement this in the retail customer segment up to 100,000 kWh/a with its existing "rigid" standard load profiles, some adjustments to the regulatory framework are required. Four essential points of special importance are:

1. Smart Meter Roll Out: The metering and control systems represent the central technical enabler for the utilisation of flexibilities. It must therefore be ensured that this new infrastructure meets the requirements of new products and that it can be used by them. From our point of view, these minimum criteria include, amongst others:

    a. Installation upon customer request

    b. Short-term, multiple, intra-day data transmission and change of switching handling, as well as possibility of implementation of, e.g. flexible pricing tariffs.

    c. Capability to balancing relevant ¼-hr values and informational minute values or similar, for example, for algorithm calculations.

    The roll out of smart meter in Germany will begin in 2017.

The implementation of new balancing procedures: To be able to provide new and/or flexible rates, the existing balancing procedures (based on standard load profiles) must be replaced by a new balancing procedure. In this respect, EnBW has developed the counter response procedure sufficiently so that it could be implemented in principle [EnBW 1]. In this context, only minor adjustments must be made to the market processes. There will be a new balancing procedure for customers with a smart meter in 2017, which will be based on ¼-h values.

Determination of the degrees of freedom in the distribution network (quotas): In order not to jeopardise the stability of the system within the distribution networks, the degrees of freedom or free capacities of individual power supplies must be known. The determination and monitoring of the degrees of freedom is associated with significant effort. This implies acceptance of costs in the context of incentive regulation.

The structure of the §14a: The current German legislative framework for energy (EnWG) includes an authorisation to issue regulations, which allows a separate network fee for interruptible consumer systems with network-enabling behaviour. The EnBW views the approach presented here as a possible replacement of current procedures used by DSOs to control electric heating devices.

In this context and the context of the structure of electricity market design, a need arises to redesign the incentives to increase flexibility potential as is properly described in the so-called "Weißbuch" from the BMWi regarding electricity market design.

With the energy approaches detailed within the pilot project, the expectations of customers regarding integration within the energy transition could be met. In addition, the current debate regarding the necessity of the use of flexibilities was supported and promoted by the pilot project with practical.

Bibliography

[BDEW 1] BDEW Bundesverband der Energie- und Wasserwirtschaft e.V.; Diskussionspapier Smart Grids Ampelkonzept – Ausgestaltung der gelben Phase; Berlin, 10th March 2015

[BMWi 1] Bundesministeriums für Wirtschaft und Energie; Ein Strommarkt für die Energiewende (Grünbuch); October 2014

[BMWi 2] Bundesministeriums für Wirtschaft und Energie; Entwicklung der erneuerbaren Energien in Deutschland im Jahr 2014; February 2015

[BNetzA 1] Bundesnetzagentur und Bundeskartellamt; Monitoringbericht 2014; 14th November 2014

[dena 1] Deutsche Energie-Agentur GmbH; dena Verteilnetzstudie – Ausbau- und Innovationsbedarf der Stromverteilnetze in Deutschland bis 2030; 11th December 2012

[EnBW 1] Dorn, Hans-Joachim; Reiter, Bernd; Warnatz, Michael; Wiechmann, Holger; Leitermann, Meinrad; Neues kosteneffizientes Bilanzierungsverfahren auf Basis der Zählerstandsgangmessung; EW spec. ed. (no. 7243) from 2014 (Jg. 113), issue 8, p. 86-89

[Schäfers 1] Schäfers, Hans; "Smart Grids – Demand Side Management Intelligente Vernetzung von Stadtinfrastrukturen"; Hochschule für Angewandte Wissenschaften Hamburg Center for Demand Side Integration (C4DSI); Frankfurt a. M.; 21st May 2014

[VDE 1] Apel, Rolf et al; ETG-Task Force Demand Side Management; VDE-Studie: Demand Side Integration – Lastverschiebungs-potenziale in Deutschland (Gesamttext); June 2012

[Wiechmann 1] Wiechmann, Holger; Neuartige Betriebsführungsstrategien für unterbrechbare Verbrauchseinrichtungen; Universitätsverlag Karlsruhe; 17th September 2008

Decentralized grid integration of electric vehicles

Dr. Gunnar Bärwaldt, Dr. L. Hofmann, Volkswagen AG

This manuscript was not available on completion of this publication.
Thank you for your understanding.

OVERALL ENERGY SYSTEMS II

Integration of renewables and electric vehicles into the smart grid – innovative energy management strategies and implementation

Yutaka Ota

1 Introduction

Electric vehicles (EVs) have important roles not only as the sustainable mobility but also as the distributed energy storage in the smart grid. Vehicle-to-Grid (V2G), which enable the EV to supply power and energy for the grid, requires the bi-directional power flow control and the two-way communication with utilities.

In this research, innovative energy management strategies of multiple EVs simultaneously considering vehicle user convenience, realistic structure of communication and control handling the multiple EVs, and harmonization to the distribution and transmission grid. The proposed strategies are being validated by a unique power, control, communication hardware-in-the-loop-simulation (HILS) implementing communication and control of these strategies into a commercialized EV, an EV supply equipment (EVSE), and a power grid real-time simulator. Both dynamical behaviour of the distributed battery and inverter interface and its system-wide effect to the power grid is evaluated in a laboratory test bed.

2 Electric Vehicles Energy Management Strategies

2.1 Overview

Overview of the energy management strategies of the multiple EVs is summarized in Fig. 1. In case huge number of EVs are simultaneously charged just after arriving at their residences, it is concerned that charging power might cause overload at the evening on the distribution transformers or feeder lines. Scheduled charging, timing of EV charge is sifted from evening to the morning just before the departure, could satisfies both vehicle user convenience and opportunity of EV demand response.

After managing the scheduled charging, each EV provides charge and discharge capability for the power grid. Renewable powered charging of the EVs is expected in the distribution grid level because the residential photovoltaic generations and the EV charging facilities are being installed into the residence at the same time. EV aggregators perform important role for realizing distribution grid-wide smart charging and solar powered charging.

Wind power generations interconnected to the power grid requires more and more flexibility of other generation resources. If enough time for the next drive, the EV quickly supplies additional charge and discharge cycles according to the grid frequency detected at the plug-in power outlet. Then the EV user obtains revenue for the frequency response, as a kind of grid-scale ancillary services [1], [2].

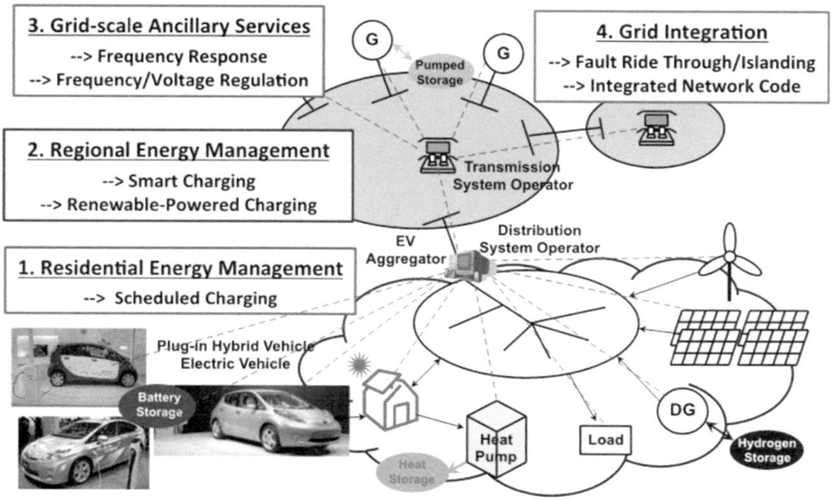

Fig. 1. Overview of Energy Management Strategies by Electric Vehicles.

2.2 Power Demand for Electric Vehicle Charging

Power demand for the EV charging is estimated based on a traffic survey by Japanese Ministry of Land, Infrastructure, Transport and Tourism. Vehicle behaviour of 17902 passenger cars is shown in Fig. 2(a). Averaged driving time and distance in the survey day is about 65[min] and 23.3[km], and the characteristics of typical commuter cars between residence and workplace are appeared.

Fig. 2(b) indicates power demand for the EV charging. It was estimated by assuming all vehicles were the EV with 20[kWh] battery, and 3[kW] chargers were equipped in all residences and workplaces. Immediate charging of the EVs in their residences just after arrival causes evening peak demand, and their peak demand can be shifted to the morning by the scheduled charging just before their departure, it is also called as timer charging. If the EVs charge up their driving energy from not only the residences but also the workplaces, power demand for the scheduled charging is divided to in the early evening at the workplaces and in the late morning at the residences.

So-called "valley filling" algorithm from the evening to the next morning is expected for mitigating the EV charging impact on the distribution feeder. On the other hands, capability of charging and discharging control is maximized in the immediate charging because the SOC (State-Of-Charge) of the EV is maintained as middle range.

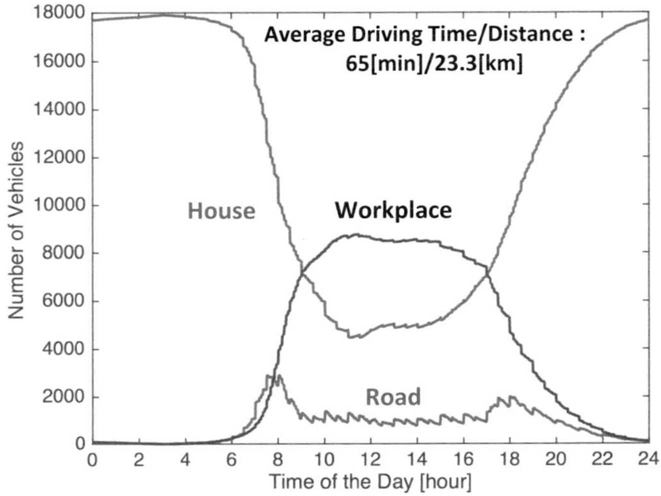

(a) Location of passenger cars

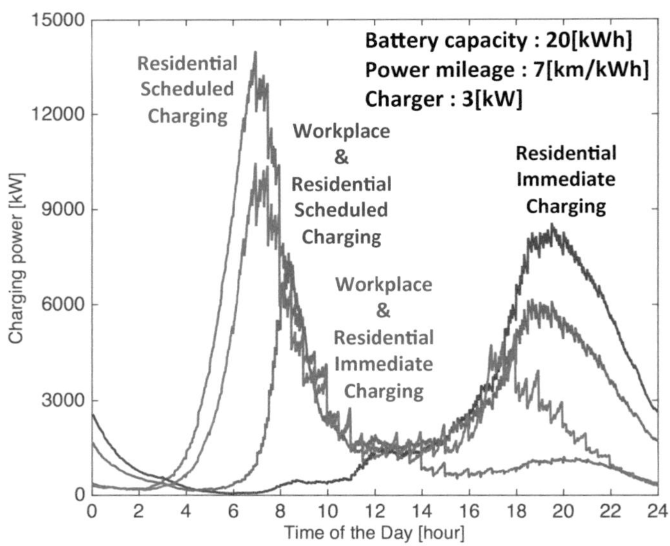

(b) Power demand for EV charging

Fig. 2. Estimation of EV power consumption based on traffic survey.

2.3 Solar Powered Charging in Distribution Feeder

Fig. 3. shows a 6600V distribution feeder model, in which 14 high-voltage customer and 479 residences, 200V, are connected to the feeder. It is assumed that all residences equip both 2.7[kW] rooftop PV (PhotoVoltaic) generations and the EV. Power flow at feeder top node and node voltage at feeder end node is shown in Fig. 4(a) and (b), respectively. Evening peak load is effectively managed by the scheduled charging of the EVs. On the other hands, reverse power flows from the PVs reduce daytime power demand, and cause voltage rise at feeder end.

An EV aggregator gathers information about the departure schedule and the SOC of each EV, and then dispatches the Vehicle-to-Vehicle (V2V) power exchange [3] signals to idling EVs in order to offset the EV charging power. Then the EV aggregator is able to bit driving energy of the EVs from the solar power generations in the distribution feeder. Solar powered charging for multiple EVs is realized by coordination of the charge and discharge dispatch signals from the EV aggregator to the idling EVs, as shown in Fig. 4.

The distribution grid-wide solar powered charging for the multiple EVs is trying to be distributed control by using the synchrophasor angle at the pick-up point of the distribution feeder as supervised information. The synchrophasor angles are also useful for detecting grid-interconnection or islanding state of the distribution sections. Therefore, the active islanding and reclose operations considering the EVs can be investigated.

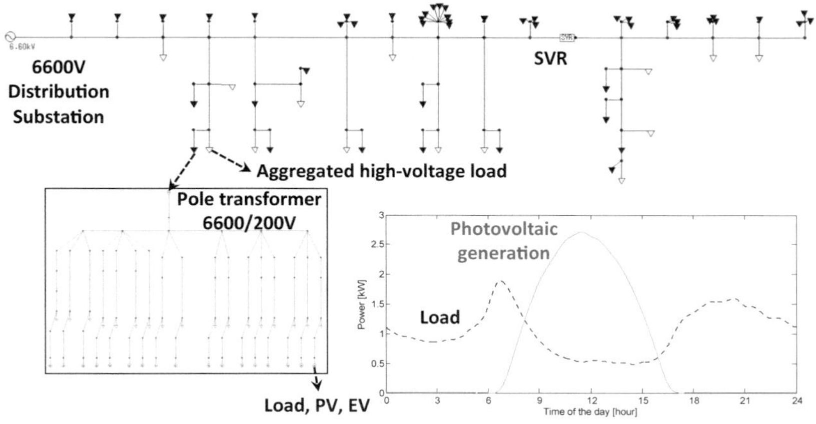

Fig. 3. Distribution feeder model with photovoltaics and electric vehicles.

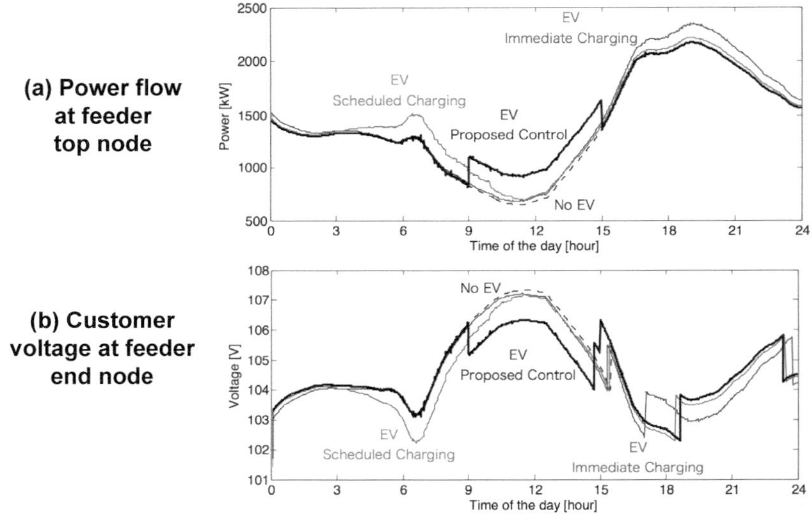

Fig. 4. Power flow and voltage calculations on distribution feeder model.

2.4 Ancillary Services for Renewable Integration

The battery energy storage and inverter system on the EV enable to supply frequency response for the power grid. Such a power control is called as the V2G (Vehicle-to-Grid) [2]. Effect of the V2G from the multiple EVs was evaluated on the Japanese power grid model with large-scale wind power generations, as shown in Fig. 5. In case of sudden power shortage on the wind power generations, quick frequency responses from the EVs contribute to improve transients of frequency, and they don't affect the conventional load frequency control by the thermal power generators, as shown in Fig. 6.

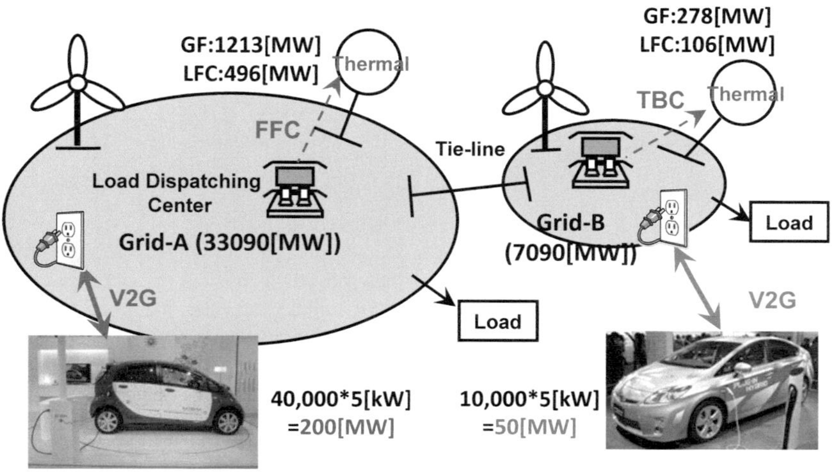

Fig. 5. Japan western-50Hz power grid model with wind power generations and electric vehicles.

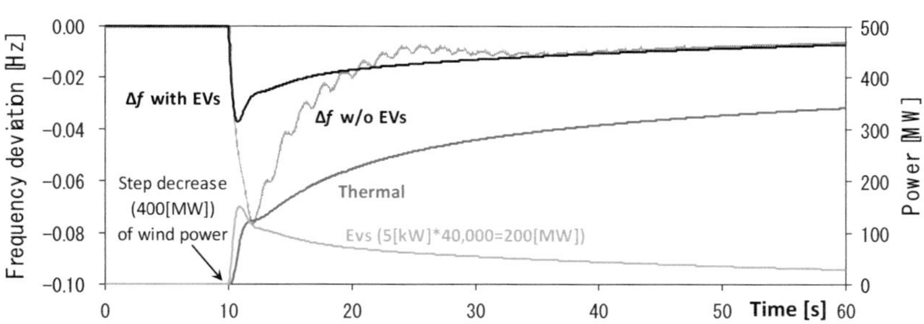

Fig. 6. Effect of electric vehicles frequency responses on grid frequency.

3 Implementation to Laboratory Hardware-In-the-Loop-Simulation Test Bed

3.1 System Configuration

System configuration of the laboratory test bed is shown in Fig. 7. Two EV and EVSE (Electric Vehicle Supply Equipment) systems are connected by VPN (Virtual Private Network) between campuses. The EVSE transforms DC 330[V] to single-phase AC 200[V], and seamlessly control charging and and/or V2G power against the power grid. The rapid prototype controller integrates three communication links between the components, and determines the power set point taking the detected grid frequency, the battery SOC, and the remote signals from the EV aggregator into account. The power grid digital simulator generates real-time control signals from the power grid model, in which large-scale renewable energy sources, multiple EVs, and generators are assumed. The proposed energy management strategies are implemented to the EV/EVSE/HILS test bed.

3.2 Test Results [4]

Step responses of the system are evaluated under the stepwise frequency change generated by the power grid simulator. Fig. 8. shows the instantaneous values of AC voltage and AC current, the grid frequency, active power output, and the battery SOC. The active power at charge is defined as positive value. The frequency change was detected with a 0.2[s] delay, and then the active power was controlled in response to the detected frequency change with a communication dead time and a delay of the power conversion.

The EV/EVSE is directly interconnected with the power grid, in which the grid frequency fluctuates as a result of the supply and demand imbalance of the actual power grid. Experimental results of the scheduled charging and the V2G are shown in Fig. 9. Just after standing up the system, the EV was charged up to 50[%] readying for the emergency departure. After the EV continued to supply the charge and discharge cycles for the frequency response, the EV was charged up to the destination SOC (90[%]).

Total settling time of the system was within one second including time lags caused by the frequency detection, communication, and response of the power conversion and control. It was enough to follow the frequency change. The seamless mode switch was also realized by the system. The power conversion loss by the EVSE was estimated by subtracting the DC power from the AC power. In the scheduled charging mode, the averaged loss was around 100[W]. In the V2G mode, the loss was slightly increased to about 200[W] because efficiency at charge was lower than that at discharge.

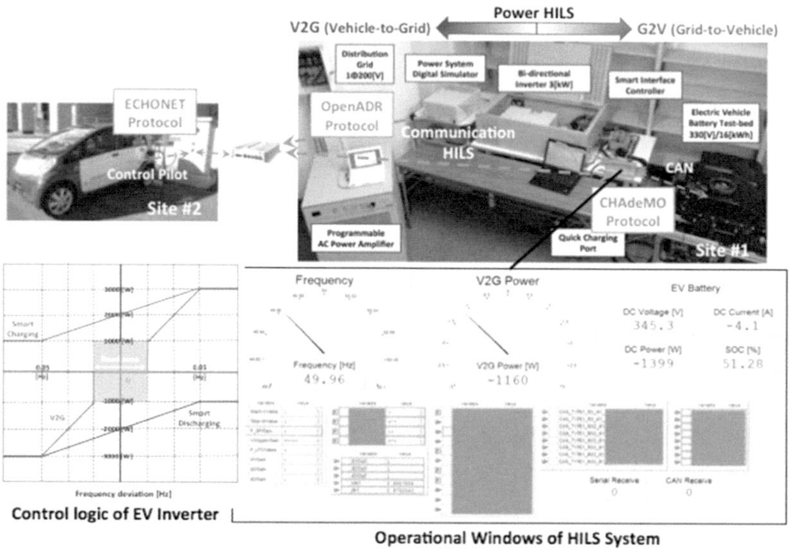

Fig. 7. System configuration of EV/EVSE/HILS test bed.

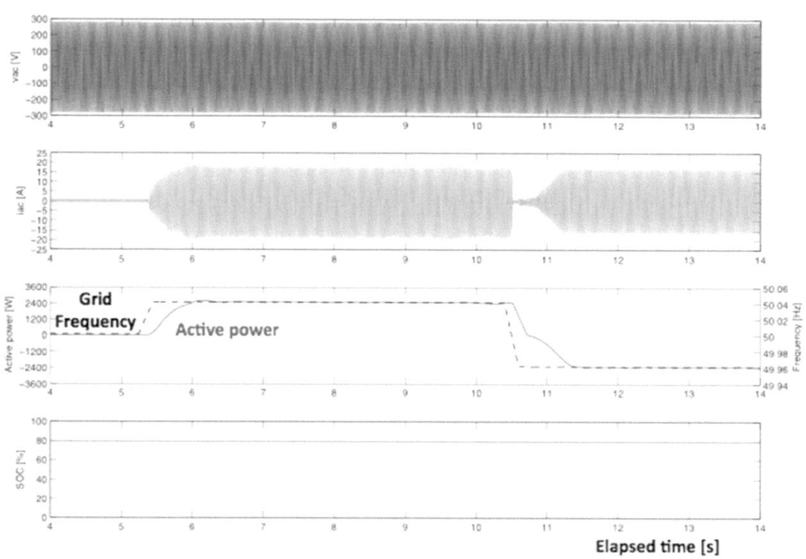

Fig. 8. Experimental results of step response.

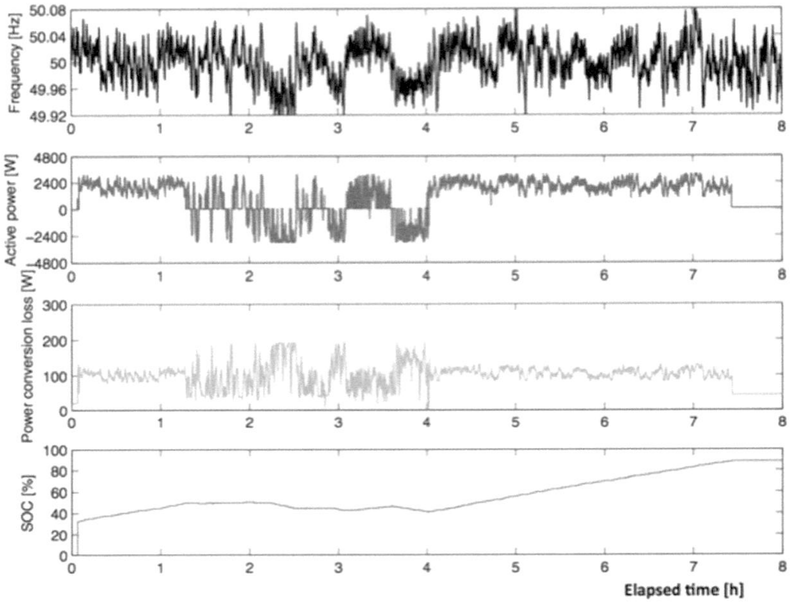

Fig. 9. Experimental results of scheduled charging and V2G.

4 Summary

In this research, the innovative "Smart Grid and E-Mobility" strategies can be realized in a laboratory test environment by use of the actual EV/EVSE and their communication and control interface. Experiences and the knowledge would be applicable to the power grid or microgrid operation and control under large-scale penetration of intermittent renewable energy sources.

It is known that there are market issues, such as the ancillary service and capacity market involving the distributed generations and energy storage devices. I'd like to challenge the market integration and the power grid issues from the experiences on the closed-loop experiment between the distributed energy storage devices and the power grid control in this research.

As a viewpoint of smart grid standardization, there are different approaches on the V2G capable connector and protocol between the EV and the EVSE. The V2G communication interface between the EVSE and the utility is also discussed. I'm conducting EV/EVSE/HILS setup integrating advanced sensing devices, such as ancillary service metering devices, IED (Intelligent Electric Device) to deliver control signals to the distributed devices, synchrophasor measurement devices, and some energy storage devices with smart inverters, as shown in Fig. 10.

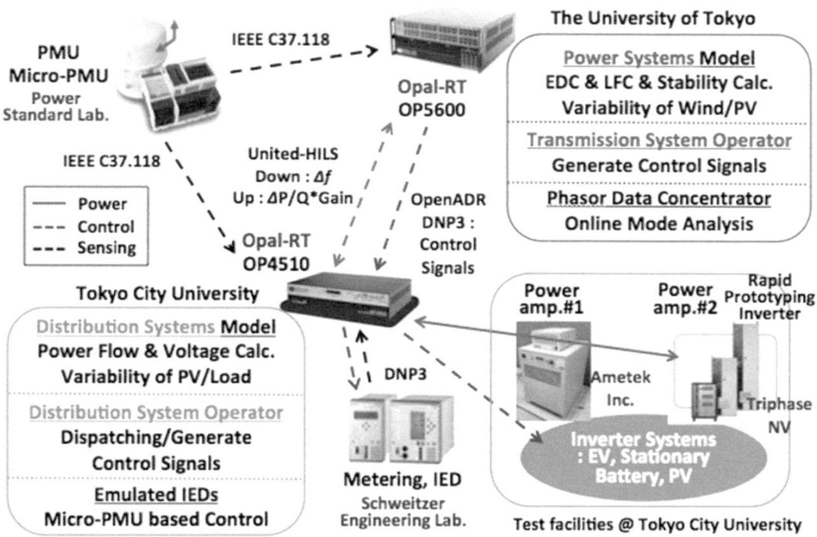

Fig. 10. Sensing, communication, and control HILS.

Acknowledgements

This research was supported by CREST from Japan Science and Technology Agency, Grants-in-Aid for Scientific Researches from Japanese Ministry of Science and Technology, and the Power Academy. Authors gratefully acknowledge their support.

References

[1] K. Shimizu, T. Masuta, Y. Ota, and A. Yokoyama, "A New Load Frequency Control Method in Power System using Vehicle-to-Grid System Considering Users' Convenience", Proc. Power System Computation Conference (PSCC), pp.1-7 (2011-08)

[2] Y. Ota, H. Taniguchi, T. Nakajima, K. M. Liyanage, J. Baba, and A. Yokoyama, "Autonomous Distributed V2G (Vehicle-to-Grid) Satisfying Scheduled Charging", IEEE Trans. Smart Grid, Vol.3, Issue.1, pp.559-564 (2012-03)

[3] Y. Ota, H. Taniguchi, H. Suzuki, J. Baba, and A. Yokoyama, "Aggregated Storage Strategy of Electric Vehicles Combining Scheduled Charging and V2G", Proc. IEEE PES Innovative Smart Grid Technologies Conference, pp.1-5 (2014-02)

[4] Y. Ota, H. Taniguchi, J. Baba, and A. Yokoyama, "Implementation of Autonomous Distributed V2G to Electric Vehicle and DC Charging System", Electric Power Systems Research, Vol.120, pp.177-183 (2015-03)

Grid optimization through electric vehicles in a cross-system comparison

Florian Samweber

Steffen Fattler

Simon Köppl

Abstract

The necessary extensive reconstruction of the grid infrastructure is a technical challenge. Its realization is caught in the field of tension between the economical and socio-political framework. Under the heading "Merit Order Netz-Ausbau 2030" (Merit Order Network-Expansion 2030 – MONA 2030), a comparison of measures and technologies for grid optimization is currently carried out, especially concerning the increased feed-in of renewable energies and the additional existence of new loads (such as electric vehicles) in the transmission and distribution grids.

Previous studies have shown that voltage problems in distribution grids only occur in a few hours per year. However, the grids have to be dimensioned for those few hours and therefore; an expensive construction of lines, controllable local grid transformers and other components is taking place at the moment. An intelligent charge management of electric vehicles that would only be necessary for a few hours, would not only reduce the constructional effort, but could also reduce costs. Additional positive effects of electric mobility in comparison to other grid optimizing measures can reside in the increased acceptance among society and positive ecological impacts.

The aim is to compare the different state-of-the-art options in a clearly arranged and striking manner. Especially for grid operators, but also for (local) energy suppliers and systems providers, alternatives regarding an useful added value to the grid planning should be demonstrated.

This paper provides an overview on currently existing grid optimizing measures. Additionally a classification of all grid optimizing measures concerning more than 60 relevant criteria have been performed by applying jj the morphological box method by Fritz Zwicky. An analysis of electric mobility as a grid optimizing measure helps to understand the concept of the proposed holistic comparison.

1 Context

The expansion of renewable energy in Germany has increased over the last years due to the goals of the German government and the hereby resulting EEG payments for renewable energy. Especially, the expansion of photovoltaic (PV) systems in the years from 2009 till 2013 resulted in the usage of the full rooftop potential for certain grid areas. This effect causes a transgression of the permitted voltage band and in some cases an equipment overload for these areas. Therefore, an additional expansion of photovoltaic solar energy results in higher costs for grid optimization for the distribution system operators.

1.1 Energy transition (Ecology, CO_2 Emission, ...)

In the light of climate change the German government decided, already before the UN climate conference COP 21 in Paris 2015, that the reduction of CO_2 emission in Germany is one of their major political goals. This transformation process is combined with a simultaneous decrease in nuclear energy usage, which is a particular characteristic for Germany among other European countries. The energy transition is supposed to result in a minimal ecological impact caused by energy production and consumption. Furthermore, main framework conditions are a minimal financial load for the economy and the citizen, as well as additional criteria such as energy supply security or noise pollution. The individual goals can be found in the expanded target triangle of energy policy. (Compare Figure 1-1) /FFE-25 15/.

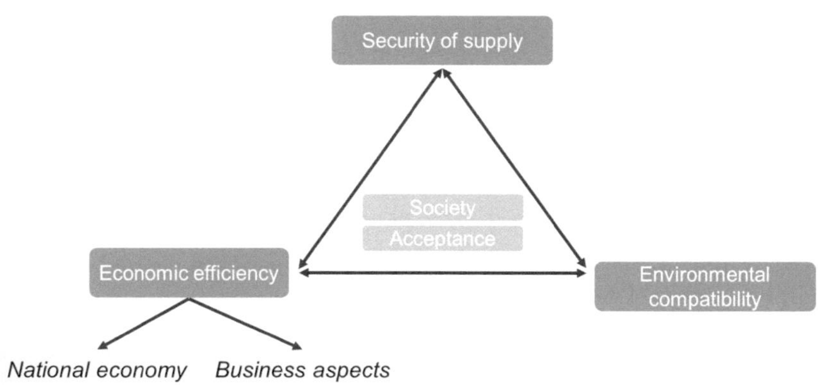

Figure 1-1: Expanded target triangle of energy policy

On the other hand, the expansion of renewable energy as a method for CO_2 reduction creates the need for measures which ensure a temporal and spatial decoupling of power generation and consumption.

Especially, the integration of photovoltaic energy into the low voltage grid system in rural areas causes "problems". Since more than 98 % of the PV facilities currently installed feed into the German low voltage grid, the main focus of grid optimization for increased PV feed-in has to lie in these grid areas.

1.2 Electrification

A goal oriented reduction of the CO_2 emission has to be combined with an increasing interconnection between the producer and consumer sectors. This way excess energy from the electricity sector can be used in the form of heat for industrial processes or private heating. Another option is a CO_2 friendly charging of the future fleet of electric vehicles.

The expecteded electrification can cause a temporal demand for high peak loads adaption in some grid areas. This modification of the grid system causes high costs. According to /DENA-14 12/ the investment costs for the German distribution grid will be between 27.5 – 42.5 Billion Euro till the year 2030, depending on the scenario.

The approach introduced in this work, which was developed in the framework of the "MONA 2030" project, compares measures that are able to reduce the costs of the necessary low-voltage grid expansion.

1.3 Goals of the MONA project

The power grid's requirements of extensive reconstruction and expansion are technical challenges. Their implementation must strike a balance between economic and socio-political conditions. Under the heading "Merit order grid expansion 2030" (MONA 2030), a comparison of measures and technologies to relieve the power grid takes place. This comparison takes into account a high proportion of renewable energy generation fed into the transmission and distribution grid.

The project MONA 2030 is supported within the scope of the research initiative into power grids by the Federal Ministry for Economic Affairs and Energy (BMWi). In addition to the BMWi, 16 industrial partners from the energy and automotive industry support the research project – both financially and through the provision of data and individual practical experience.

In the project four main focus points are going to be investigated in more detail for a holistic assessment:

Scenario analysis:

Future developments and parent influences will be presented and examined in the first stage of the project. This will create a foundation for a long-term planning horizon for grid expansion, as well as for the establishment of investment decisions. Besides the current trend (reference scenario), this approach will develop several other scenarios using a consistency analysis. In contrast to comparable studies these scenarios specifically differ in the various influence opportunities for the grid actors. In the scope of the project, actors are defined as transmission and distribution system operators as well as prosumers.

Basic information:

There are various grid and supply structures. However, to provide a common basis for the evaluation of measures regarding the relief of the network, grid data will be evaluated and clustered into basic grid topologies. Subsequently grid types will be classified using regional generation and consumption profiles of very high resolution, in order to cover a variety of different grid loads.

Comparison of grid optimizing measures:

As a basis for the decision making process of stakeholders (i.e. utilities and grid operators), both existing and new, innovative methods of network relief will be scientifically analyzed. This analysis will be accompanied by practical suggestions for application and comparative studies. Furthermore the comparison of measures will include techno-economic parameters as well as societal and ecological aspects. In conclusion, an analysis above the primary grid optimization will be deducted, which does not only include a safe network operation, but the overall added value for the energy system.

Cost-optimized ranking of the measures (merit order)

To fulfil the objective of a foresighted, holistic network, all planning measures regarding grid relief must be designed efficiently and cost-effectively. The aim is to provide an appropriate basis for an assessment that is transferable and scalable. Hence, recommendations for integrated grid planning in Germany can be derived.

The results of MONA 2030 will not only provide a basis for political discussion. Moreover, they will support the stakeholders affected by network expansion with selecting appropriate measures for the realization of a cost-optimized security of supply of electricity grids.

2 Methodology: holistic approach through a morphological box

The complexity of the ranking process of suitable measures for the reconstruction of the grid infrastructure will be simplified using a comprehensible and value-free classification, which includes all the relevant aspects (economic efficiency, technology, ecology, public acceptance, etc.) for the interdisciplinary comparison. As a suitable tool for this classified comparison the morphological box was chosen.

2.1 Introduction to the morphological box

The morphological box was introduced by the astrophysicist Fritz Zwicky and describes a heuristic method for the decomposition of a complex task into distinctive sub-aspects. These sub-aspects are structured into superordinate categories and subordinated criteria. This methodology facilitates an objective illustration of a vast number of evaluation criteria for the examined measures. The result is a matrix, which contains all the possible manifestations of the criteria. The methodology of the morphological box is characterized by transparency, reproducibility and expandability. /FFE-39 15/

2.2 Evaluation of grid optimizing measures through the morphological box

A six step approach was chosen for the evaluation of the grip optimizing measures (gom) in the morphological box. This will ensure an objective evaluation and a goal oriented comparison.

(1) Definition and generalization of the problem

In the first step the problem will be defined and generalized. A clear formulation and definition of the question will help to reduce the necessary effort for the evaluation of the available options.

(2) Determination of relevant criteria

The categories „technology“, „economy and politics“, „society and environment“ and „secondary added value and contribution to operational management“ are defined in order to map the various effects of the gom in the context of the cross-system comparison.

The category „technology“ includes all criteria of the individual measures, which describe the physical parameter and technical properties like operation site and efficiency. The category „economy and politics“ summarizes all measures, which determine operational parameters, the economic environment and the regulatory framework conditions, like investments and lifetime. The category „society and environment“ contains all criteria of the individual measures, which affect the acceptance among the population and the surrounding environment. Examples are noise pollution, interventions into landscape and land usage. The category „secondary added value and contribution to operational management“ is comprised of the total secondary added value for the energy system, which is created by the measures, and their contribution to the operational management. This includes for example the effect on the operational management and supply of operation reserves.

(3) Identification and classification into relevant criteria

At this stage each category is assigned with a variety of criteria, which are identifiable and independently from each other. They are determined through expert surveys and literature research.

(4) Determination and limitation of the criteria manifestations

There exist discrete and continuous manifestations for every criterion whose number can be vast and are often in different units. The manifestations will be specified as discrete and concrete steps or clear units, in order to evaluate all measures in a consistent manner.

(5) Completion of the morphological box for concrete gom

At this stage the morphological box is completed for each chosen gom using data sheets, empirical values, calculations, surveys and further methods. The overall goal is an objective and complete classification of all criteria.

Figure 2-1 shows the combination of the manifestations exemplary for the two gom „electric vehicles for network relief“ and „variable local grid transformer“ with different colours. Both gom can be described by a clear and discrete manifestation. Only the criterion „investment“ is an exception as it is described by concrete numerical values.

technology	**application site**	substation	transformer	junction box	line	house connection point
	required ICT	no ICT required	low	medium	high	
economy and politics	**stakeholder**	producer	grid operator	aggregator	consumer	sales
	Investitionen	very low: only additional ICT	low	medium	high	very high
society and environment	**structural change**	new construction	expansion of existing facilities	optimizing of existing facilities		
	noise emissions	no emissions	low	medium	high	
secondary added value and contribution to operational management	**resilience against future changes**	low	medium	high		
	provision of reactive power	yes, capacitive or reactive	yes, inductive	yes, capacitive	no provision possible	

Figure 2-1: Extract of a morphological box filled out for the two gom „electric vehicles for network relief" (orange) and „adjustable distribution grid transformer" (green) /FFE-12 15/, /FFE-23 15/.

(6) Evaluation and comparison of the individual solutions

At the end, the reader is able to identify the optimal gom for each criterion. An adaption of the individual grid planning to personal requests can be achieved through a weighting of the criteria and categories. A possible objective and scientific method is for example the cost-benefit analysis.

3 Holistic analysis of grid optimizing measures

As an introduction the basic assumptions under which the grid optimizing measure (gom) "electric vehicles (EV) for grid relief" are evaluated and described. These evaluations are performed under the current boundary conditions and assumptions for the year 2030.

Technology description

Concerning the evaluations of charge management for EV only private charging points will be considered. Public charging points in indoor and outdoor car parks will be neglected since the energy charged there only accounts to about one percent of the total energy charged /EEXP-01 11/. Furthermore, it is assumed that the required charging infrastructure is already available. Currently EVs in private households are either charged with the use of a standard Schuko plug (~3.7 kW, 1-phase) or a Wallbox additionally installed. The latter is usually connected to a high-power line supporting charging capacities of up to 22 kW (3-phased) /MEN-1 14/. Concerning the charging infrastructure no significant changes are anticipated in the considered timeframe till 2030. To enable such a charge management of EVs today, a ripple-control receiver, a so-called control-box and an additional metering point for billing-purposes would have to be installed. In 2030 only an additional control box, realizing the communication between the Smart Meter Gateway and the charge management of the EV, will have to be installed since the Smart-Meter-Rollout is planned to be completed by then /BRG-01 15/. In the project MONA 2030 the influence of four charge management types on nine standardized grid topologies will be evaluated on the basis of low-voltage grid simulations. Besides an uncontrolled direct charging and a voltage dependent charge management, charging optimized for self-consumption or economic efficiency, will be implemented in the simulation and evaluated. The results of these simulations will be published in the final report of the project in spring of 2017.

The battery capacity of EVs is a critical factor concerning the potential of the charge management for grid optimization. Most of the currently available EVs have a battery capacity of about 22 kWh resulting in an average range of about 150-190 km (compare to: /BMW-02 14/). By estimations from /FFE-4 16/ the average EV will have a battery capacity of about 50 kWh in 2030. Besides the battery capacity of the EVs the total penetration is of vital importance for their potential for grid optimization. According to the "Kraftfahrzeugbundesamt" on 1st of January 2015 18.948 EVs and 107.754 Plug-In-Hybrid-Electric-Vehicles (PHEV) were registered in Germany by 2015 /KBA-03 15/. In /FFE-04 16/ various studies dealing with the future penetration of EVs were compared leading to an estimate of about 3.3 million vehicles in 2030 (EV and PHEV).

In the following chapter the gom „electric vehicles for grid relief" will be described with the use various criteria represented in a Morphological box.

3.1 Technology

In the category „technology" 22 criteria where identified to describe the technical properties of the gom „electric vehicles for grid relief". The most significant ones are examined in this chapter.

As described in the beginning of the chapter, the majority of the charging processes will be performed at home. Most of the grid connection points will therefore be connected to the **low-voltage grid**. Depending on the penetration rate of EVs the gom offers the **possibility of a multi-usage** at various locations in the grid area. Currently the use of charge management of EVs for grid optimization is not yet commercially available, but various **pilot projects for research on possible charge management types** are carried out /BMU-09 11/. In all likelihood the Smart Meter Rollout will be completed by 2030, allowing an economic and easy implementation of charge management systems in private households. Therefore, it can be assumed that this development phase will be completed as well as charge management of EVs will widely be commercialized (at least theoretically). Due to the relatively low penetration of EVs even in 2030 the **efficiency as a gom** is to be considered as relatively low as well (see /FFE-04 16/). In general the charging process of EVs can **dynamically be adjusted by adjusting the charging capacity** and is able to follow **steep gradients**. Many battery storage systems already fulfil the very strict prequalification requirements of the German "Primärregelleistung" (primary control reserve), highlighting their technical potential. Apart from that it remains to be seen if the comparatively high expense of a dynamic charge management of a large number of EVs can turn out to be an economically feasible option in comparison to the other gom. Limited by the EVs battery capacity and the times in which the EVs are not "at home", relatively **low call duration** has to be assumed as well as a **daytime dependent availability**. To attain a certain potential as a gom a large number of EVs will have to be pooled. Therefore, the ICT demand is to be considered as relatively high since data monitoring and control technology will have to be implemented at various locations in the grid area. On the other hand this broad distribution of technical unit's leads to a **high resilience against disturbances**, since the malfunction of one single unit can be well compensated by the large number of charging points.

3.2 Economics and politics

The current regulatory framework does not define the charge management of EVs as a grid optimizing measure. Hence, no corresponding renumeration model is implemented. According to /HILL-01 14/ charging infrastructure and EVs are not defined as grid operating resources, but as units of consumption. Therefore the grid operator is not allowed to directly influence the operation of these units. Solely § 14a of the Energiewirtschaftsgesetz (EnWG, German Energy Act) currently allows the grid operator to use certain units of consumption for means of grid relief, explicitly including EVs. According to this paragraph, these interruptible consumers are renumerated with a reduced network charge if they have an additional metering point installed and can be used by the grid operator for grid relieving purposes /ENWG-01 15/. In practice this passage is currently not applied for EVs. To be used as a gom in the future, various concepts are conceivable for the charge management of EVs. For example an **aggregator** could offer special tariffs for EV owners allowing him to actively influence the charge management of the included EVs as in means of grid relief. Another possible concept would be the individual and decentralized charge management, optimizing the self-consumption of a household for example in combination with a PV system. The implementation of these concepts would enable the end consumer to smooth out his peak load, effectively supporting voltage stability in the grid. In both cases the use as a gom would represent a **potential source of income for the owner of the EV**. The conducted charge management could either be renumerated by the aggregator, the responsible grid operator or – in case of the optimization of self-consumption – be incentivized by the economic savings the self-consumption generates in comparison to a conventional electricity purchase. Under the current regulatory framework the grid operator is not permitted to directly influence the operation of charge points installed in private households (except for EnWG §14a). Therefore, corresponding adjustments would be needed. Generally, the **technical potential of EVs as a gom** is limited by the absolute penetration of EVs, which also results in **geographical differences** concerning the potential.

3.3 Ecology

The types of charge management discussed are to be seen as a mere **optimization of existing technical units**. Therefore, any kind of **effect on the environment** is negligible. Also the **technical risk level** and the cohering **degree of damage** is to be estimated as low.

3.4 Society

Today as well as in 2030 the participation of end users for charge management in private households will most probably be voluntary and/or incentivized. Hence an active involvement of the society in the decision-making process is not deemed to be necessary. For an estimation of the public acceptance of the various gom in MONA a new method for ranking order assessment was developed. Based on the results of a survey of all the industrial partners involved in the project the method facilitates a formal comparison of all the goms. In this study the gom "electric vehicles for grid relief" will be compared to the well-established gom "conventional grid expansion". For this purpose the results of the survey are summarized in an evaluation matrix and settled with the corresponding weighting factors (see Table 3-1).

Table 3-1: Ranking order assessment of the grid optimizing measures

		conventional grid expansion		electric vehicles for grid relief	
attribute	weighting in %	score*	weighted score	score*	weighted score
external effect	6%	2	12	4	18
costs	10%	2	20	5	50
landscape	10%	1	10	5	50
added value	10%	2	20	3	30
technological risk	1%	2	2	4	4
association	12%	1	12	5	60
contribution to the "Energiewende"	8%	1	8	5	40
fairness	1%	1	1	4	4
trust	14%	1	14	4	60
prior experience	9%	5	45	1	9
public involvement	2%	2	4	4	8
desirability	1%	1	1	4	4
voluntariness	7%	2	14	4	28
procedural design	6%	2	12	4	24
public opinion	4%	1	4	4	4
total assessment		26	179	60	393
rank order		2		1	

*score from 1 (very low) to 5 (very high)

According to the results of the rank order assessment the gom "electric vehicles for grid relief" reaches a higher score than the "conventional grid expansion". Therefore, it can be stated that the charging management of EVs has a higher degree of fulfillment of factors relevant for acceptance. The results show that the gom "electric vehicles for grid relief" will be accepted in the society with a high degree of probability. For a detailed evaluation of topic see /HED-01 15/.

4 Overview on potential grid optimizing measures

In order to discuss the role of electric vehicles as a grid optimizing measure, the project MONA 2030 studies all relevant options. Within the analysis of 19 relevant European studies, an online survey with 125 participants among stakeholders such as economists, researchers, politicians, associations and students and one expert interview, 14 grid optimizing measures have been identified (compare Figure 4-1). All of them are able to integrate more power into the electric grid.

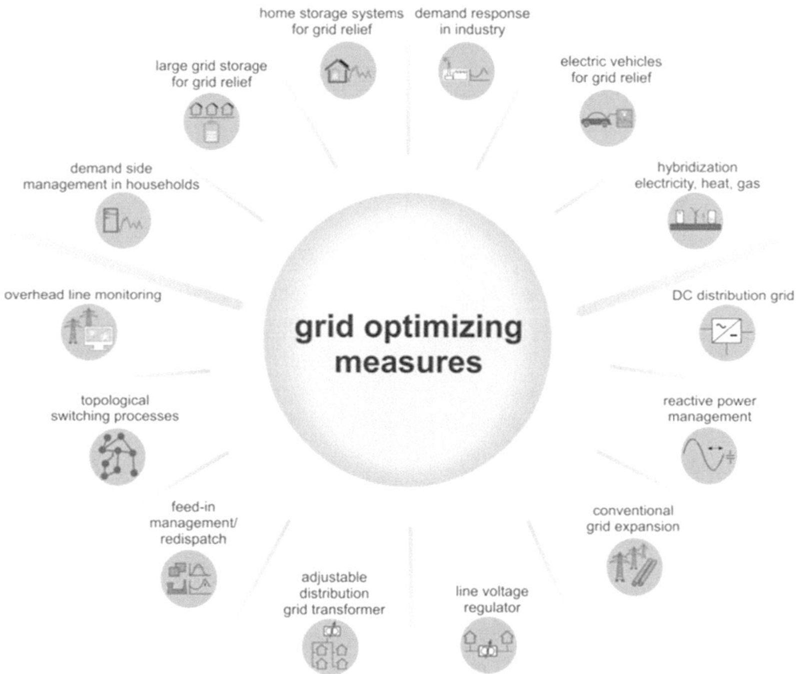

Figure 4-1: Potential grid optimizing measures identified within the project MONA 2030

As not all grid optimizing measures shown in Figure 4-1 are well established yet, a short description of all measures is given here.

Conventional grid expansion depicts the strengthening of the transmission and distribution grid infrastructure by additional construction and upgrading of the power transmission system (e.g. overhead power lines, underground cables, substations, HVDC). Especially in the distribution grid, conventional grid expansion is the most established grid optimizing measure.

The grid optimizing measure **overhead line monitoring** defines the weather-related overloading of overhead lines to reach higher transmission capacities under cold and windy weather conditions. The realization of the measure requires an examination and – if necessary – an adjustment of circuit components. /TEN-03 12/

Feed-in management describes the temporary reduction of renewable feed-in power. The responsible grid operator is therefore – under certain circumstances and if grid capacities are insufficient – able to limit the preferential feed-in of these plants, but is obliged to reimburse the plant operator for the lost energy /BNETZA-03 14/.

Redispatch are interventions in the generation capacity of power plants in order to avoid overload of line sections. All plants at all network levels with a net rated active power of 10 MW or bigger are obliged to do redispatch /BNETZA-08 15/.

The measure **demand response in industry** defines a temporary and predictable change of energy demand (flexible loads) in reaction to price fluctuations, incentive payments due to high prices on the wholesale market or a threat to system stability. /FFE-42 10/

The grid optimizing measure **hybridization electricity**, heat, gas describes the conversion of electrical energy to heat or the usage of electrical energy to produce methane in order to avoid inadmissible power values and so to guarantee grid stability. The measure includes new consumers such as heat pumps that can be controlled flexibly.

Reactive power management characterizes reactive power compensation systems or the directed control of inverter-paired plants in order to reduce or control reactive power within the electricity grid. In the future, decentralized generation plants will not only have to consume inductive reactive power, but also provide reactive power to support voltage stability.

Adjustable distribution grid transformers can increase the reception capacity of distribution networks. This can be achieved by the installation of a switchable on-load tap-changer which results in a dynamically controllable output voltage of the transformer.

A **line voltage regulator** keeps the output voltage of the electricity supply within the defined limits and can therefore increase the reception capacity of the grid. In comparison to the controllable distribution grid transformer, the line voltage regulator is installed within a distribution grid feeder and can locally avoid voltage range deviations.

Topological switching operations can increase the quality of supply. Due to the switching operation, the grid topology can be ideally adjusted based on actual or predicted feed-in and load scenarios.

The grid optimizing measure "**stationary battery storage for grid relief**" offers new possibilities for the integration of renewable energies into the power grid. Electricity from different local generation plants that cannot be used directly is stored and therefore increases grid stability and decreases transmission losses. High loads can be fed through strategic discharge of the storage.

Home storage systems for grid relief can regulate – due to the connection to the main power supply – the feed-in of electricity from renewable energy sources into the grid and therefore contribute to grid relief. Through reduction of the active power feed-in when consumer load is low and decentralized generation is high, the voltage can be kept within the defined limits. Another advantage depicts the reduction of active power losses in the distribution grid. /ECOW-01 14/

It has to be considered that the main motivation for installing a home or a large grid storage system is an increasing self-consumption and energy-self-sufficiency by prosumers or communities. For a comprehensive grid optimizing effect, new business use cases for storage systems to stabilize the grid have to be developed.

The grid optimizing measure **DC distribution grid** requires a conversion of the distribution grids from AC to DC. The measure can lead to a cost reduction of the distribution grid infrastructure, higher transformer efficiencies, lower losses in lines and voltage converters and reduced material intensity /FEN-01 15/. Security risks and a time- and cost-intensive conversion from today's AC-based infrastructure have to be taken into account.

Flexible loads in households can contribute to grid stability and balance fluctuations of the residual load through demand side management in households. Through DSM, less back-up capacities are needed and the need for feed-in management can be reduced. Additionally, the need for energy storage can be reduced and it can contribute to more flexibility within the power system.

5 Conclusion

The cross system comparision of electric vehicles with other grid optimizing measures points out strengths and weaknesses of this measure. In total, 14 measures which are able to solve future challenges in the electric grid have been considered as part of this analysis.

Using the upcoming smart meter infrastructure, electric vehicles can benefit from low additional costs compared to conventional grid optimizing measures. Intelligent charging algorithms can help to increase the self sufficiency of supply area. The permanent availability of this measure for grid relief is crucial for distribution grid operators. Therefore, there can be a preference to conventional grid assets instead of innovative goms. Nevertheless, intelligent charging strategies can help to minimize bottlenecks in the electric grid.

The project MONA 2030 compares all options in different scenarios and typical distribution grids. Final results will be available by the end of 2017.

6 References

BMU-09 11 Gesteuertes Laden V2.0. Berlin: Bundesministerium für Umwelt, Naturschutz und Reaktorsicherheit (BMU), 2011

BMW-02 14 BMW i3: Technische Daten in: https://www.bmw.de/de/neufahrzeuge/bmw-i/i3/2013/techdata.html. München: BMW AG, 2014

BNETZA-03 14 Leitfaden zum EEG-Einspeisemanagement – Abschaltrangfolge, Berechnung von Entschädigungszahlungen und Auswirkungen auf die Netzentgelte, Version 2.1. Bonn: Bundesnetzagentur (BNetzA), 2014

BNETZA-08 15 Hinweise zur Durchführung von Redispatchmaßnahmen vor dem Hintergrund der Urteile des OLG Düsseldorf vom 28.04.2015. Bonn: Bundesnetzagentur (BNetzA), 2015

BRG-01 15 Gesetzentwurf der Bundesregierung – Entwurf eines Gesetzes zur Digitalisierung der Energiewende. Berlin: Bundesregierung, 2015

DENA-14 12 dena-Verteilnetzstudie – Ausbau- und Innovationsbedarf der Stromverteilnetze in Deutschland bis 2030. Berlin: Deutsche Energie-Agentur GmbH (dena), 2012

ECOW-01 14 Häni, Martin: Grundlagenbericht Stromnetze Thurgau – mit Blick auf eine verstärkte dezentrale Stromproduktion. Bürglen, Schweiz: EcoWatt AG, 2014

EEXP-01 11 Fischer, Norbert: Warum Ladesäulen auf sehr lange Zeit kein ertragreiches Geschäft sein werden. München: emobility-experts, 2011

ENWG-01 15 Gesetz über die Elektrizitäts- und Gasversorgung (Energiewirtschaftsgesetz – EnWG). Berlin: Bundesrepublik Deutschland, 2015

FEN-01 15 Flexible Elektrische Netze Konsortium (FEN): Warum Gleichspannung? in: http://www.fen.rwth-aachen.de/cms/FEN/Der-Forschungscampus-FEN/Ueber-das-FEN-Konsortium/~hkju/Ueber-das-FEN-Konsortium/. Aachen: FEN RWTH Aachen, 2015

FFE-04 16 Pellinger, Christoph; Schmid, Tobias; et al.: Merit Order der Energiespeicherung im Jahr 2030 – Hauptbericht. München: Forschungsstelle für Energiewirtschaft e.V. (FfE), 2016

FFE-12 15 Nobis, Philipp; Samweber, Florian; Fischhaber, Sebastian: PV-Eigenverbrauch mit Elektrofahrzeugen und Hausspeichersystemen – Veröffentlichung im Rahmen der 9. Internationalen Energiewirtschaftstagung an der TU Wien 2015 (IEWT). München: Forschungsstelle für Energiewirtschaft e.V. (FfE), 2015

FFE-23 15 Wohlschlager, Daniela; Köppl, Simon: Kostenanalyse Netzstabilisierungsmaßnahmen – Stand Mai 2015 – Kostenparameter für Komponenten ausgewählter netzstabilisierender Maßnahmen im Niederspannungsnetz. München: Forschungsstelle für Energiewirtschaft e.V. (FfE), 2015

FFE-25 15 Pellinger, Christoph; Samweber, Florian: Merit Order der Energiespeicherung und des Netzausbaus. In: FfE Fachtagung 2015, Tagungsband und Vorträge. München, 2015, ISBN: 978-3-941802-28-5

FFE-39 15 Köppl, Simon; Wachinger, Kristin; Samweber, Florian: Der Morphologische Kasten: Eine Vergleichsgrundlage für Netzoptimierende Maßnahmen in: et – Energiewirtschaftliche Tagesfragen 65. Jg. (2015) Heft 10. Essen: etv Energieverlag GmbH, 2015

FFE-42 10 von Roon, Serafin; Gobmaier, Thomas: Demand Response in der Industrie – Status und Potenziale in Deutschland. München: Forschungsstelle für Energiewirtschaft e.V. (FfE), 2010

HED-01 15 Hedeler, Barbara: Gesellschaftliche Akzeptanz netzoptimierender Maßnahmen – Konzeption eines Vergleichsdesigns zur Einbindung gesellschaftlicher Akzeptanz in das Forschungsprojekt Merit Order Netzausbau 2030. Masterarbeit. Herausgegeben durch die Technische Universität München und die Universität für Bodenkultur Wien, betreut durch die Forschungsstelle für Energiewirtschaft e.V.: München: 2015

HILL-01 14 Hiller, Thomas; Bodach, Mirko; Castor, Walter: Praxishandbuch Stromverteilungsnetze – Technische und wirtschaftliche Betriebsführung. Würzburg: , 2014

KBA-03 15 Bestand an Pkw am 1. Januar 2015 nach ausgewählten Kraftstoffarten in: http://www.kba.de/DE/Statistik/Fahrzeuge/Bestand/Umwelt/2014_b_umwelt_dusl_absolut.html (abgerufen 25.02.2016). Berlin: Kraftfahrt-Bundesamt, 2015

MEN-01 14 Weinert, Wolfgang: Ladestecker-Standards in der IEC-Norm: EC 62196, IEC 61851 – Presseinformation. Kirchhundem: MENNEKES Elektrotechnik, 2014

TEN-03 12 Schmale, Michael: Witterungsabhängiger Freileitungsbetrieb bei der TenneT TSO GmbH – FGE-Kolloquium Sommersemester 2012. Aachen: TenneT, 2012

Rollout e-Mobility – the next big challenge for network operations and network planning

Dr. Armin Gaul, Project Leader, RWE International SE, Germany;
Christian Czajkowski, RWE International SE, Germany;
Stephan Voit, RWE International SE, Germany

Stefan Übermasser, AIT, Austria

This paper is based on the work of the PlanGridEV consortium:

Introduction

The mass roll-out of electro-mobility, in conjunction with the increased production of electricity derived from renewable sources, is one of the major opportunities to reduce CO_2 over the next decade. The efficiency of electric engines is superior to that of combustion engines and the renewable energy sources used to power them are naturally CO_2-free. Additionally, the fact that electric cars use batteries takes care of the issue of energy storage, helping to overcome the problem of renewable sources only being available when the sun is shining or the wind is blowing. The battery in the car, in combination with an energy management system that can monitor the availability of renewable energy, is a major opportunity. [1]

It is estimated that by 2025, about five million electric vehicles (EVs) will be on the road all over the world, the majority of these in the EU. European climate policies aim to significantly reduce CO_2-emissions from transport sector by 60 % by 2050 and to reduce the use of "conventionally-fuelled" vehicles in urban transport by 50 %. With technological advances and new mobility concepts, as well as a steadily rising oil price, it can be assumed that electro-mobility in Europe will increase. Simultaneously, the feed-in from DER is also expected to dramatically increase in order to meet the goals of EU and national climate protection policy. Therefore it can be expected that distribution grids will need to significantly increase their hosting capacity to accommodate fluctuating supply and mobile loads, in such as EVs. [1]

The challenge of these two growing movements is the requirement for large scale interaction between energy supply and demand. While this is a well-known balance in need of management in industrial production plants, it has never been applied at grid level. Moreover, the fact that the controlled loads (EVs in this case) are moving around and are only connected to the grid from time to time has never previously been taken into account. To date, there have been no grid planning tools designed to meet these opportunities and challenges. [1]

Large-scale EV introduction will moreover only be successful if the expectations of customers are met and current constraints are overcome. This requires further technological development and the intense cooperation of all stakeholders, including OEMs, DSOs and energy service providers, for example in operating a sufficient (fast) charging network infrastructure. [1]

The new hurdle for grid planning is that although the grid extension needed to provide the energy for EVs is actually rather small, the power need, especially for fast charging, is substantial. In the end, it's not only the delivery of the energy that is the problem but the power itself. The delivery of the power might depend on the charge scheme in use: fast versus grid-friendly. Challenges arising from this are line loading issues and voltage as well as the necessary introduction of ICT in the low voltage (LV) and medium voltage (MV) grids. [1]

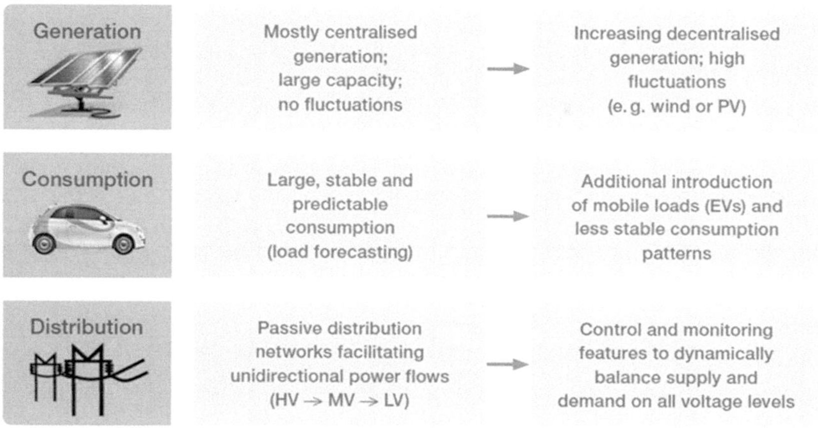

Figure 1: Dramatically changing environment demands new grid planning rules [1]

PlanGridEV approach

To overcome these challenges, the EC funded project 'PlanGridEV' has developed a totally new planning approach for distribution grids, implemented via a prototype tool. Instead of simply considering the estimated peak load of the grid, as with traditional distribution grid planning, the 'PlanGridEV' approach additionally considers the controllability of the loads in conjunction with the estimated generation from renewable sources. Within 'PlanGridEV' synthetic models for the major components of car movement and charging, wind generation, PV generation and storage have been developed and integrated. The new planning tool takes into account not only one peak load day but stochastic behaviour over a several month planning horizon. Additionally an ICT communication model has been developed and integrated to derive the necessary ICT components for a certain implementation. [1]

Based on a set of scenarios and use cases extensive simulations have been performed that prove that there are major benefits with a combined smart grid and smart charging approach. Depending on the set-up, the peak load could be reduced by half compared to uncontrolled charging scenarios. Furthermore the usage of renewable energy could be increased without interfering with transportation needs. This clearly shows that in an all-electric world the smart grid approach will play a major role. [1]

Flexibility

One of the major tasks of a DSO is to deliver as much energy (power and profile) as a customer requests for a location. If the current grid situation does not fit customers' requirements a DSO has to enhance the capacity of the grid. Traditional grids are designed to fulfil peak load which occurs in residential areas usually around noon and in the evening. This is the reason why the most transformer have enough spare capacity to charge EVs, but this capacity is available at other times. Figure 2 shows the free transformer capacity related to one feeder out of a simulation of 'PlanGridEV'. At 1 pm it would be only allowed to charge one EV with 11 kW (corresponds in this scenario to an EV penetration of 7%). On the other hand 37% of the transformers capacity is unused and there is enough spare capacity to provide the energy for more than one EV. Using this spare capacity by available flexibility could avoid grid extension.

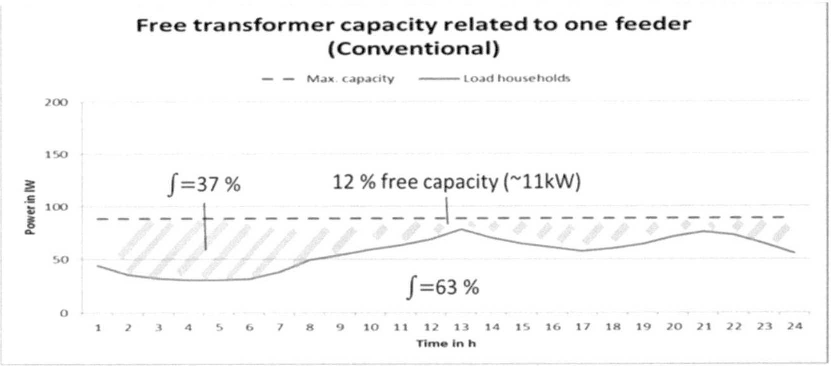

Figure 2: Free energy capacity at one feeder [3]

Uncontrolled charging in practice

Figure 3 shows the analysis of 450,000 charge sessions at public charge poles in 2014. Based on the load duration curve a coincidence factor was calculated. In about 300 hours all connected cars were simultaneously charging at full power and the maximum energy of the used charge poles have been consumed for charging. That equals a coincidence factor of 1 and means at an interesting and large charging location it is likely that exactly this happens.

Therefore we need to calculate with an coincidence factor of 1. Then the curve is decreasing fast and after around 3400 hours of the year (less than half a year) none of the charging processes took more than 40% of the possible energy (equals coincidence factor of 0.4). [2]

Figure 3: Coincidence factor at public charge poles [2]

The graph in Figure 4 shows the average charging behavior within the sample of charging processes in Germany on around 400,000 real charging sessions at public charge points operated by RWE within one year. So no night time recharge at home is integrated in these datasets. As an assumption the charging energy is taken as constant over the loading time. A clear daily profile is visible which has to be separated between working day and weekend. On a typical working day the charging peak is reached by a tough gradient at about 9 o'clock in the morning and a second peak is reached at 1 pm. [2]

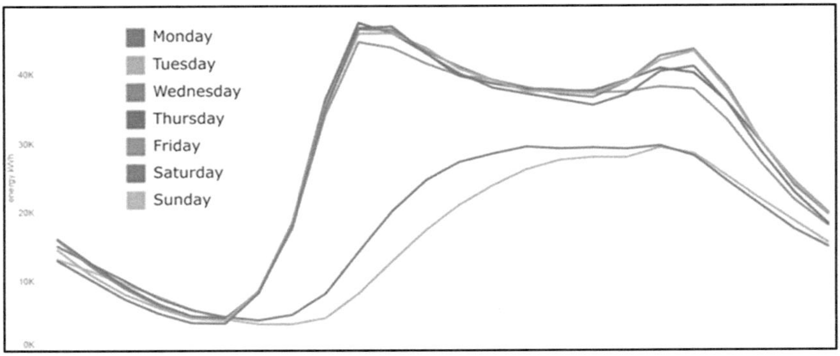

Figure 4: Energy consumption over the day [2]

Furthermore studies ([4],[5],[6]) assume that most EVs will be recharged in the evening hours after work when arriving at home (private charging). This approach follows also a paper which analysis probabilistic load models to derive load profiles for EVs. [6] With the EV penetration rate and the size of the EV chargers a standard load profile for charging EVs can be derived from the arriving time based on the MID Study (Mobility in Germany). The peak occurs at 6 pm for 43.5 kW charger and at 7 pm for 3.7 kW charger.

Consequences of uncontrolled charging

In case of uncontrolled charging a diversity factor of 1 must be assumed for each charging location. As Figure 3 shows a diversity factor of 1 for all connected EVs is not only in theory but also practically possible and will occur in more than ten days a year (around 320 hours). A grid planner has to dimension the grid for the worst case since it cannot be ensured that not a very high number of EVs will arrive at the same time and start immediately with charging at full power. If this is one-to-one applicable to a single LV feeder needs to be studied further, but it cannot be excluded and must therefore be taken into account from planning perspective. [2]

With respect to the LV grid and private and public charging it means for the grid planner that they have to consider every charging point at the feeder, because he has to assume a diversity factor of 1. There is a real probability that all connection points are covered by EVs and that they charge with maximum power. This power must be scheduled for the feeder and the transformer in the substation. [2]

Integrating EVs at work or at semi-public areas will lead to a power increase in the amount of the maximum power that can be achieved by loading all cars at the same time (diversity factor of 1 since uncontrolled charging). This power increase has to be paid by the enterprise or the location's operator and is fully effective in grid planning and dimensioning the feeder and the transformer. [2]

With respect to the MV grid (transformer in the substation) the added EVs will increase the power and a diversity factor of 1 must be assumed again from the perspective of a DSO, but the substation is the property of the enterprise or the location's operator and will not influence the public grid as MV lines are not the limiting factor for charging EVs. The owner of the substation has to dimension the transformer in a suitable way to enable the simultaneous charging at all charging points. This applies to all locations connected to the MV grid. [2]

Flexibility offered in practice

The correlations between driving patterns and charging locations and the resulting time to charge EVs is clearly evident. When we talk about private charging at home the peak is located in the time frame of peak load in a residential area in the evening. Public charging corresponds to the peak around noon or occurs in the morning.

So it is necessary to create charging occasions where less power is needed and where there is more time for charging. To still provide the flexibility that customers ask for, charging will occur while the car is planned to be parked anyway. [2]

The "Grid Charging Flexibility Value" GCFV is defined as

$$GCFV = \frac{Duration_{Charge}}{Duration_{Charging\ Session}}$$

If $\text{Duration}_{\text{Charge}} = \text{Duration}_{\text{Charging Session}} \rightarrow \text{GCFV} = 1$ there is no flexibility offered, and therefore the charging cannot be delayed or shifted. The closer GCFV gets to zero, the more flexibility an EV offers to define a charging profile. The GCFV of about 230,000 real charging sessions at charge points operated by RWE is shown in Figure 5. 62 % offer a GCFV smaller than 0.5, the duration of these charging sessions is at least twice as long as the minimum time needed for charging. A lot of the charging sessions may offer flexibility for scheduling charging according to grid requirements or to energy services and the customer would be able to provide flexibility without giving up their needs. [2]

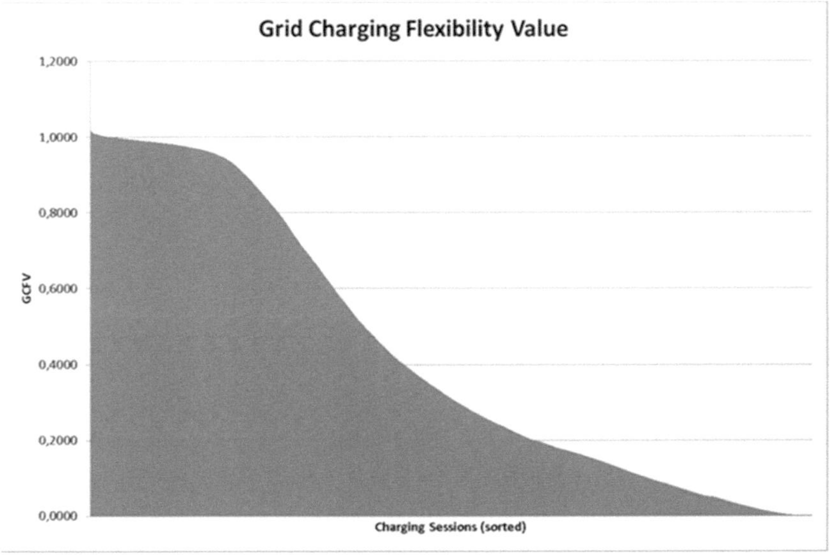

Figure 5: Grid Charging Flexibility Value [2]

Figure 6 shows the average ratio between charging and connection time for different sizes of chargers based on these real charging sessions in comparison to real charging sessions at work (fleet).

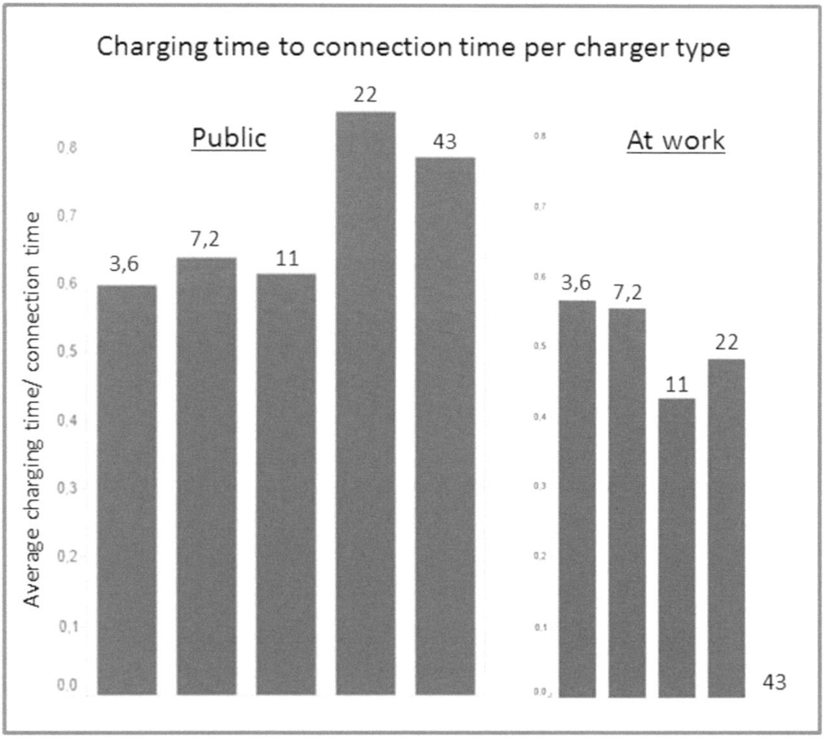

Figure 6: Average charging time/ connection time by charger [2]

The comparison of the theoretically available flexibility of the charging process shows a significant difference between the charger types 11 kW and smaller and the more powerful ones. For smaller charger the utilization of the connection time is on average about 60% while for the lager chargers it goes up to 95%. That means with larger chargers you see more the behaviour of filling at a gas station. [2] This is due to the fact that the driving patterns differ to those with the smaller chargers. The comparison to charging at work shows, that EVs with a charging power of 11 kW or 22 kW can offer more flexibility. The smaller charger remain almost unchanged. In general a higher charging power as well as bigger batteries will increase the flexibility.

Opportunities of controlled charging

In case of controlled charging a diversity factor of 1 for all connected EVs can be avoided. Studies ([5],[6]) show that the diversity factor can be reduced by enabling controlled charging.

With respect to the LV grid and charging at home it means, that the connection point can be neglected for grid planning. As the car is parked overnight and enables enough flexibility to charge the car at times with a low penetration of the feeder, more charging points are possible at one feeder. Furthermore a higher charging power may be allowed to be connected to the LV grid since there is flexibility available and a higher charging power improves also the diversity factor. [2]

Public parking spaces are also connected to the LV grid but due to the consumers behaviour they are not as well suited for controlled charging as private charging. As shorter parking durations can be assumed and time for flexibility is needed, controlled charging at public parking spaces maybe can avoid short peak loads at the LV grid. Higher charging power would improve this flexibility. [2] As Figure 7 shows public charging is necessary to avoid peaks and the saturation of the transformer at noon in case of a high EV penetration. In addition DER hosting capacity and the possibility to charge a higher number of EVs in parallel will increase.

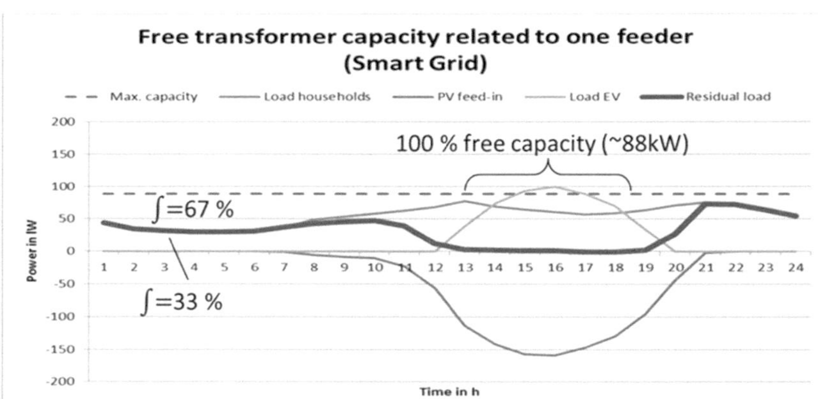

Figure 7: Smart charging increase power capacity [3]

Charging at work and semi-public charging have to be differentiated too. Where charging at work offers enough time for flexibility and to use controlled charging, the parking times at semi-public spaces have to be assumed shorter. So semi-public spaces are similar to public charging and benefits from a higher charging power. Controlled charging at work leads to the possibility to integrate more EVs (fleet management) without increasing the peak power. [2]

By nature planning is confronted to handle with vague and future related assumptions for long term (decades). Energy management systems at households and working premises which are connected to the DSO to be informed on the real-time grid status and actual limitations can help to improve the planning process. For the DSOs these systems must be reliable that the grid limits will be respected. Energy management systems could then be are aware of clear and strict power limits per customer to manage (automatically) the power and energy demand itself. On the other hand this requires a lot of effort to bring these information together.[2]

Project results

'PlanGridEV' has been able to show with the prototype tool, that a new grid planning approach is both feasible and leads to desired results. It could be proved in the simulations and in the test beds that the new approach leads to less expensive grids while providing all the energy the customer needs. In conclusion, we have found a way in which a full roll-out of electric vehicles is feasible. The implementation requires that regulation and the market develop accordingly, with respect to the new functionality that will be available in the coming years. [1]

An evaluation of the benefits of smart charging in conjunction with renewable generation has been carried out. Already in the European Alternative Fuels Infrastructure Directive 94/2014 the need for smart charging was addressed in 2014. [8][8] This directive has to be implemented in the member states till the end of 2016. With the results of the studies carried out by 'PlanGridEV' there is several points that need to be established to create a framework based on the directive that paves the way to a mass-rollout of e-Mobility.

The major finding is that smart charging is the only way for a mass-rollout of e-Mobility that does not generate massive costs to reinforce the grids.

The 'PlanGridEV' results prove that the additional energy needed to supply the cars can be easily provided by the existing energy system while the power demand of uncontrolled charging would result in massive grid investments.

A similar effect has been recognized with the mass integration of PV especially in the German grid. In the beginning control of PV generation has not been taken into account and this has been resulted in major grid investments, so finally a directive was enacted that peak generation curtailment and control mechanisms have to be implemented also in existing DER systems.

To establish the framework two things are necessary.

1. The proposed architecture needs to be available.
2. A market model needs to be established that enables smart charging business.

'PlanGridEV' has proven that the technology to build the proposed framework is available at a technology readiness level (TRL) of 6 which means that the proposed control schemes could be verified within the test beds of the four participating DSOs. Additionally we did laboratory tests showing that all EVs on the market are generally able to fulfil the needs of the proposed approach but most of them are still lacking a high level control protocol even though that this protocol is available and standardised in ISO 15118.

Conclusions and stakeholder recommendations

A market model to generate financial benefit out of these above described technical capabilities is still missing. 'PlanGridEV' has described in detail how it needs to be designed and has analysed that all the necessary communication and interaction between the different market players.

To implement the market model two more steps are needed.

1. All market players need to include the same implementation in their product.
2. The remuneration scheme that empowers smart charging has to be implemented.

To achieve the proposed results the stakeholders have to implement the following steps.

EV Manufactures

Full implementation of an advanced standard (e.g. ISO 15118 and IEC 61851-1) to support controlled charging in all new vehicles, which allows multiple charging activities per day and modulated charging.

Enable charging at rates up to 22kW AC that can be controlled via appropriate protocols (e.g. ISO 15118) to provide access to the full potential of flexibility of EVs.

Implement 3-phase AC chargers for all EVs that draw more than 3.6 kW to avoid unbalance in distribution networks.

Implement smart charging mechanism for home charging that can benefit from load demand management for example to increase the usage of local PV generation.

Implement smart charge control mechanisms for DC charging (if DC charging is implemented in the EV) since there are significant cost–effect benefits between completely uncontrolled charging patterns and controlled ones.

Increase the battery capacity. This reduces the need to recharge the EV during peak time (passive benefit) and increases the capacity of flexibility (active benefit).

Charging Infrastructure Operators and Manufactures

Ensure that all private and public charging infrastructure and charging equipment is capable of controlling EVs via ISO 15118 and IEC 61851-1.

Integrate interfaces to the DSOs to guarantee that grid friendly operation is possible.

Integrate interfaces to home and company energy management systems for increasing the usage of renewables and peak shaving to lower peak power prices (e.g. on company premises).

Figure 8: Smart cards currently needed to do test drives in Europe and charge at different charge pole operators [1]

Integrate certified and automated metering to be prepared for business models that give benefits to the costumers if they participate in certain control schemes (e.g. Operating reserve)

DSOs

Support grid friendly infrastructure by:

Establishing a framework for reduced grid fees for charge poles / charge pole operators that have fully implemented control schemes in accordance with ISO 15118 and IEC 61851-1 and a working and qualified interface for interactions with the DSO. Reduced grid fees could be:

- Reduced prices for kilowatt-hours.
- Reduced prices for peak power.
- Reduced grid connection fees.

Introduce a prequalification process for suitable equipment and operational principles.

Integrate interfaces between grid control systems and the charging infrastructure (e.g. through the operator of the charging infrastructure) to guarantee that grid friendly operation is possible.

Integrate interfaces to home and company energy management systems for an increased use of renewables and power demand optimisation.

Define limits of power demand respecting the local grid

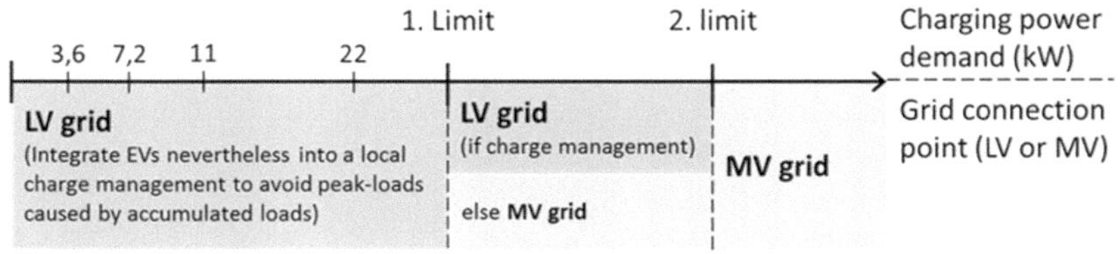

Figure 9: Defined limits of power demand respecting the local grid [9]

Avoid to support uncontrolled charging by:

Calculate the adequate grid costs for the connection of charging infrastructure which is not qualified for controlled charging and therefore has to be rated as a customer with a connection that guarantees the peak power of the charging infrastructure any time of the day.

Strictly claim for adherence of the existing grid connection specifications (e.g. limits for non-symmetric loads).

Regulators

Ensure that regulation demands the necessary implementations from all stakeholders as explained above to pave the way of a mass-rollout of EVs.

Especially update the regulatory framework to ensure

- That also the DSO can benefit by investing in smart grid technologies instead of traditional grid extension.
- That the customer can benefit from offering his flexibility by allowing to charge his car grid friendly.

Since the avoided grid reinforcements play a major role in the business case for a mass-rollout of EVs, it's necessary to empower the DSOs to have an active role in the management of renewables and loads (charging EVs) by providing grid friendly demand management services to the market.

Allow the DSOs to calculate the above proposed change in the grid fees to promote grid friendly charging infrastructure.

Politicians

Claim for realization of the Directive 94/2014 (European Alternative Fuels Infrastructure) by implementation of controlled charging equipment and mechanisms by all stakeholders as explained above to pave the way of a mass-rollout of EVs.

Support the controlled charging mechanisms and the proposed modification of the regulatory framework.

Only provide public funding or incentives for:

- EVs that fully implement ISO 15118 and IEC 61851-1.
- Charging equipment that implement ISO 15118 and IEC 61851-1 and has an interface with the DSO.
- For charging equipment that is compliant with the necessary power demand to enable the usage of flexibility and fulfils the grid connection specifications of the DSOs.
- Charging equipment that is compliant with the necessary power demand to enable the usage of flexibility and fulfils the grid connection specifications of the DSOs.
- Charging infrastructure that is connected to grids (LV/MV) with sufficient reserve for the anticipated mass-rollout.

- Charging infrastructure if the necessary certified metering is implemented at the interface that the DSO is responsible for.
- Introduction of power tariffs, to bring incentives to customers to perform controlled charging and energy management at home locations.
- In the case of power tariffs at end customers, temporary reduced costs for having an incentive to invest in EVs

Support the proposed framework and the mechanisms that are already in charge and avoid increasing complexity by adding alternatives for available market roles (e.g. support qualified metering systems in the charging infrastructure and not mobile metering systems).

Acknowledgements

The research leading to these results has received funding from the European Union Seventh Framework Programme (FP7/2007-2013) under grant agreement No. 608957.

References

[1] Armin Gaul: PlanGridEV, Distribution grid planning and operational principles for electric vehicle mass roll-out while enabling integration of renewable distributed energy resources, Final Report, February 2016, www.plangridev.eu

[2] Armin Gaul, Daniel Feismann: PlanGridEV, Handbook/White Paper of Novel Planning Rules for EV in Smart Grids, Deliverable No: 6.3, March 2016

[3] Armin Gaul, Christian Czajkowski: PlanGridEV, OEM Workshop: How to benefit from smart charging, 19.02.2016

[4] Fraunhofer-Institut für Windenergie und Energiesystemtechnik (IWES): Wissenschaftliche Unterstützung bei der Erstellung von fahrzeugbezogenen Analysen zur Netzintegration von Elektrofahrzeugen unter Nutzung erneuerbarer Energien. Final Report. , (available online at http://www.bmub.bund.de/fileadmin/Daten_BMU/Pools/Forschungsdatenbank/fkz_um_11_96_107_elektromobilitaet_bf.pdf access in February 2016)

[5] LBD-Beratungsgesellschaft mbH: Einbindung von Zuhause Kraftwerken in Smart Grids. March 2012, (available online at http://www.lbd.de/cms/pdf-gutachten-und-studien/1203_LBD-Gutachten-Einbindung_ZHKW_in_Smart_Grid.pdf access in February 2016)

[6] J. Linssen et al.: NET ELAN – Netzintegration von Fahrzeugen mit elektrifizierten Antriebssystemen in bestehende und zukünftige Energieversorgungsstrukturen. 2012, (available online at http://www.emobilserver.de/fileadmin/user_upload/emobil%20neu/Energie%26Umwelt_150.pdf access in February 2016)

[7] Alexander Probst; Martin Braun; Stefan Tenbohlen: Probabilistic load modeling and simulation for households and electric vehicles for voltage band analysis.

[8] DIRECTIVE 2014/94/EU OF THE EUROPEAN PARLIAMENT AND OF THE COUNCIL of 22 October 2014 on the deployment of alternative fuels infrastructure

[9] Armin Gaul, Christian Czajkowski: PlanGridEV, Final Review Meeting: 21.04.2016

KEYNOTE LECTURE

Operational experience of grid-integrated vehicles with V2G in three countries

Prof. Dr. Willett Kempton,
Center for Carbon-free Power Integration,
University of Delaware,

S. Fisher, NRG EVgo;

S. Fisher, PJM Interconnection;

G. Poilasne, Nuvve Corporation, USA

This manuscript was not available on completion of this publication. Thank you for your understanding.

MIX
Papier aus verantwortungsvollen Quellen
Paper from responsible sources
FSC® C105338

If you have any concerns about our products,
you can contact us on
ProductSafety@springernature.com

In case Publisher is established outside the EU,
the EU authorized representative is:
Springer Nature Customer Service Center GmbH
Europaplatz 3, 69115 Heidelberg, Germany

Printed by Libri Plureos GmbH
in Hamburg, Germany